Dear Neil Armstrong

PURDUE STUDIES IN AERONAUTICS AND ASTRONAUTICS

James R. Hansen, Series Editor

Purdue Studies in Aeronautics and Astronautics builds on Purdue's leadership in aeronautic and astronautic engineering, as well as the historic accomplishments of many of its luminary alums. Works in the series will explore cutting-edge topics in aeronautics and astronautics enterprises, tell unique stories from the history of flight and space travel, and contemplate the future of human space exploration and colonization.

(●)

RECENT BOOKS IN THE SERIES

Dear Neil Armstrong

Letters to the
First Man from
All Mankind

James R. Hansen

Purdue University Press
West Lafayette, Indiana

Cataloging-in-Publication data is on file with the Library of Congress.

Hardback ISBN: 978-1-55753-874-1
EPUB ISBN: 978-1-61249-603-0
EPDF ISBN: 978-1-61249-602-3

Letters featured in this volume are from the Neil A. Armstrong papers in the Barron Hilton Flight and Space Exploration Archives, Purdue University Archives and Special Collections.

To all of the people who wrote letters to Neil—
it is to their respect, admiration, and fascination
with him that I dedicate this book

APOLLO 11

CONTENTS

FOREWORD

Throughout history there have been ordinary men and women who accomplished extraordinary things. Neil Armstrong was one of those. As an experimental test pilot for NASA, he flew the X-15 at Edwards Air Force Base, going higher and faster than anyone before him, flying to the edge of space. And then he transferred to the astronaut program so that he could fly in space.

I was a member of the fifth selection group of astronauts in 1966 and arrived in Houston in May of that year, just two months after the flight of Gemini 8. Neil had been the commander on that flight, which ran into trouble when the attitude control thrusters fired suddenly, causing the spacecraft to tumble. Both the Gemini 8 and crew could have been lost that day, but a coolheaded Neil got the spacecraft under control.

When I arrived that May the talk was all about how Neil had saved the flight. And since I was already excited about being assigned to the Manned Space Center, all that talk about Gemini 8 only served to get my heart going faster. My introduction to Neil was the result of our being on the same floor at the MSC, and I saw him only intermittently in my early days there. Busy with post–Gemini 8 responsibilities, Neil was not in the office much. But he was quick to introduce himself and welcome me, showing friendly support for my new assignment to the Astronaut Office. I felt honored to meet him.

There is an old saying that the ultimate test pilot is the one who, when faced with certain death, calmly reads out the instruments before crashing.

Without a doubt the ability to keep a cool head is a preeminent characteristic of great test pilots. Neil Armstrong certainly demonstrated that, as witnessed by the way he both saved Gemini 8 and landed the lunar module on July 20, 1969. During the descent to the Moon an onboard computer overload caused a series of alarms to sound. Then as the lunar module *Eagle* dropped closer to the surface, Neil saw that the area they were headed toward was rocky and knew he must find a more open area in which to land. As *Eagle* approached 500 feet he took manual control, and with lowered visibility from a layer of lunar dust kicked up by the descent engine and a dwindling fuel supply, he successfully landed the lunar module in the Sea of Tranquility.

Neil Armstrong was often described as stoic, but that was not my experience of him. Although I did not know him in an intimate, familiar way, we were friends, and we got to know one another better after the Apollo program ended. While I served as chairman of the Astronaut Scholarship Foundation and as such focused on raising money for that organization, Neil focused on raising funds for Purdue. But he was always ready to speak at ASF events, and at an anniversary celebration for Apollo 15, for which I served as the command module pilot, he showed his warmth and grace as he spoke about the importance of the CMP role during the Apollo program, giving much credit to Michael Collins, his CMP and second in command on Apollo 11.

I hope you enjoy these samplings of the many thousands of letters and well-wishes Neil received, and responded to, from people all over the world. Though they are thoroughly enjoyable in their own right, through them we gain insight into the First Man and what he meant to so many.

Al Worden
July 2019

PREFACE

Among the items the crew of Apollo 11 left on the Moon was a tiny silicon disc about the size of a half-dollar. Etched onto that disc, in microscopic lettering about one-fourth the width of a human hair, were messages of goodwill from seventy-four leaders of the world's nations, messages that had been rather hurriedly solicited by NASA (after authorization by the U.S. State Department) just weeks before the launch of Apollo 11. Also etched on the disc were four presidential statements: from then current U.S. president Richard M. Nixon and the immediate past U.S. president Lyndon B. Johnson, plus quotes from the National Aeronautics and Space Act of 1958, signed by Dwight D. Eisenhower, and John F. Kennedy's May 25, 1961, lunar landing commitment speech to Congress. The disc also contained the names of various members of the U.S. Congress who had served on committees instrumental to the achievement of the Apollo program, as well as the names of the leading NASA manned space program officials. In truth, NASA had contacted 116 countries asking for their contributions to the historic disc of goodwill messages, but the deadline for producing the disc came so fast that not all were able to get their messages back to NASA in time.

As the goodwill messages were not written expressly for Neil Armstrong, nor for his crewmates, nor for NASA, and most of them not even directly for the United States of America, it may seem unusual to use them as the basis of the introduction to this book dedicated to Armstrong's letters. But in another respect these messages are the perfect way to lead the reader into the contents, and the spirit, of this book.

For one thing, Neil Armstrong commanded the mission that took the disc to the Moon. But more than that, if not for Neil the small beat cloth package containing the disc, preserved within a thin aluminum case, might not have been left on the Moon, at least not during Apollo 11. The package was in the shoulder pocket of Buzz Aldrin, Neil's lunar module pilot and fellow Moonwalker, who had been so busy with his tasks on his EVA (extravehicular activity) that he had completely forgotten about leaving the package on the surface. Buzz was halfway up the ladder heading back into the lunar module *Eagle* when Neil, still down on the surface, called up to him, "How about that package out of your sleeve? Get that?" Buzz pulled the packet out and tossed it down onto the surface, at which point Neil gave it a nudge with his foot, straightening it out a little and getting some dust off it.

Thus, the Apollo 11 silicon disc was gifted to the Moon and, of course, rests in that place today, hopefully never to be moved (or removed) by future lunar explorers. Along with the disc in that cloth bag were a few other highly significant mementos. There was an embroidered patch from the never-launched Apollo 1 mission, in memory of fellow astronauts Roger Chaffee, Gus Grissom, and Edward White, who died when their command module caught fire during a routine test on the launch pad at Cape Kennedy in January 1967. Also in the bag were two Soviet-made medals, in honor of pioneering Soviet cosmonauts Yuri Gagarin, the first human in space, who had died in a MiG-15 accident in March 1967, and Vladimir Komarov, killed a month after Gagarin at the conclusion of his Soyuz 1 flight when his spacecraft's descent parachute failed to open. Inside there was also a gold replica of an olive branch, a symbol of peace.

All seventy-four messages of goodwill on the Apollo 11 disc are notable and merit the recollection of our global community, now fifty years since they were expressed. For the purpose of this book, and a reasonably brief preface, however, a sampling of these messages must suffice. What follows then is a suite of historic messages, resting now for half a century on our Moon's Sea of Tranquility, from our planet's seven continents (less Antarctica), sent in peace and hopeful of a better future, three each from Africa, Asia, Europe, North America, and South America, and three from Australia in company with the islands of Oceania—a total of twenty-one messages. In choosing which messages to include, I have done my best to put myself in the frame of mind of Neil Armstrong and the kind of goodwill messages that would likely have impressed him the most in July 1969 as well as today.

AFRICA

From W. V. S. Tubman, president of Liberia

"It is extraordinary, almost incredible that what was but an idea, even a little over a decade ago, seems now an attainable reality. Man's imagination, ingenuity and technology have not only impelled him to look up but has also enabled him to reach the celestial bodies.

"The journey to the Moon which these three Americans, Commander Neil Armstrong, Colonel Edwin E. Aldrin, Jr., and Lt. Colonel Michael Collins have successfully undertaken is a voyage to the celestial and these messengers of discovery represent the people of every country.

"We salute these explorers of outer space and pray for their security and safety while we admire their courage and intrepidity.

"I ask them to bear this message to the inhabitants of the Moon if they find any there. If they do not, it is my desire that this message be one of the greetings from the people of Liberia and myself to the Moon, nebulous satellite of the Earth."

From J. D. Mobutu, president of the Congo

"The government of the Democratic Republic of the Congo follows with constant attention the achievements of human genius in the conquest of space in order to make man its master. The Congolese people, its party, its government, and myself express our ardent wish to see Apollo 11 successfully accomplish the mission which is our own. May these victories which have cost man so much energy and sacrifice continue to the reinforcement of cooperation among peoples and serve peace for the greatest good of mankind. Best regards."

From Félix Houphouët-Boigny, president of the Ivory Coast

"At the moment when man's oldest dream is becoming a reality, I am very thankful for NASA's kind attention in offering me the services of the first human messenger to set foot on the Moon and carry the words of the Ivory Coast.

"I would hope that when this passenger from the sky leaves man's imprint on lunar soil, he will feel how proud we are to belong to the generation which has accomplished this feat.

"I also hope that he would tell the Moon how beautiful it is when it illuminates the nights of the Ivory Coast.

"I especially wish that he would turn towards our planet Earth and cry out how insignificant the problems which torture men are, when viewed from up there.

"May his work, descending from the sky, find in the Cosmos the force and light which will permit him to convince humanity of the beauty of progress in brotherhood and peace."

ASIA

From Indira Gandhi, prime minister of India

"On this unique occasion when man traverses outer space to set foot on Earth's nearest neighbor, Moon, I send my greetings and good wishes to the brave astronauts who have launched on this great adventure. I fervently hope that this event will usher in an era of peaceful endeavor for all mankind."

From Park Chung-hee, president of the Republic of Korea (South Korea)

"The landing on the Moon by Apollo 11 is a brilliant feat of all mankind which makes men's dreams a reality and marks a new chapter of human history. This great achievement is a result of man's constant striving for progress towards a brighter destiny. Now, realization of man's adventure into yet further reaches of space seems but a few steps away.

"On this historic occasion, we do solemnly pledge ourselves to work together on this Earth for the better world with lasting peace and prosperity for all mankind. Let us celebrate the first landing of men on the Moon, the symbol of eternal grace and the mirror of man's true heart, with a new spirit which will inspire mankind to realize the ideal of civilization in which men live in justice, freedom, and unity."

From Cevdet Sunay, president of Turkey

"I consider the landing of the men on the Moon as a sign of the beginning of a new era of which we could hardly dream until now.

"Since the start of the space explorations, the Turkish nation has most ardently wished the realization of this thousand-year old dream and followed with great hope and excitement every success in this field.

"I wish to congratulate the most heartily the heroic astronauts and the

American people, our friends and allies, for they have spared no effort in this field and also those who have contributed to the achievement of this outstanding accomplishment from which, I am sure, mankind and our civilization will benefit for peaceful purposes."

AUSTRALIA AND OCEANIA

From John Gorton, prime minister of Australia
"Australians are pleased and proud to have played a part in helping to make it possible for the first man from earth to land on the moon. This is a dramatic fulfillment of man's urge to go 'always a little further,' to explore and know the formerly unknown; to strive, to seek, and to find, and not to yield. May the high courage and the technical genius which made this achievement possible be so used in the future that mankind will live in a universe in which peace, self-expression, and the chance of dangerous adventure are available to all."

From Keith J. Holyoake, prime minister of New Zealand
"By this flight man has finally fulfilled the ambition of setting foot on another celestial body. As Prime Minister of New Zealand I hope that the realization of this dream—so long remote—will inspire all those who set their sights high and thus bring closer the dreams of peace and cooperation for all mankind."

From Seewoosagar Ramgoolam, prime minister of Mauritius
"Your bold venture commands admiration of government and people of Mauritius wish you God's speed and happy perfect landing. Safe return."

EUROPE

From Elizabeth R., queen of England
"Message from Her Majesty Queen Elizabeth II
"On behalf of the British people I salute the skill and courage which have brought man to the moon. May this endeavor increase the knowledge and well-being of mankind."

From Baudouin, king of the Belgians

"Now that, for the very first time, man will land on the moon, we consider this memorable event with wonder and respect.

"We feel admiration and confidence towards all those who have cooperated in this performance, and especially towards the three courageous men who take with them our hopes, as well as those from all nations, who were their forerunners or who will follow them in space.

"With awe we consider the power with which man has been entrusted and the duties which devolve on him.

"We are deeply conscious of our responsibility with respect to the tasks which may be open to us in the universe, but also to those which remain to be fulfilled on this earth, so to bring more justice and more happiness to mankind.

"May God help us to realize with this new step in world history better understanding between nations and a closer brotherhood between men."

From Américo de Deus Rodrigues Thomaz, president of Portugal

"The Portuguese people, discoverers of the unknown Earth in centuries past, know how to admire those who in our days explore outer space bringing mankind in contact with other worlds."

NORTH AMERICA

From Pierre Elliott Trudeau, prime minister of Canada

"Man has reached out and touched the tranquil moon. Puisse ce haut fait permettre a l'homme de redecouvrir la terre et d'u trouver la paix. (May that high accomplishment allow man to rediscover the Earth and find peace.)"

From Gustavo Díaz Ordaz, president of Mexico

"It is an honor for Mexico, with this most modest symbolic testimony, to form part of the event which for the first time takes man to a soil away from his home planet. And, in doing so, Mexico extends most enthusiastic congratulations to the dedicated, gallant astronauts and to the scientists and technicians, as well as, in a broader sense, to the American people and their Government for this undertaking that, hitherto, only had precedents in the realm of imagination.

"Mexico's very own emblem—its traditional seal—with the eagle and the serpent, already embodies the double sign inspiring man since his remote origins and which in a particular manner may be equated to coming humanity: the serpent represents flight, undaunted and far-seeing, a fearless pilgrimage which makes it possible for the legacy of the centuries to reach ever increasing circling horizons. Far from being contradictory to each other, both images are complementary and placed together reflect our temporal, earthly nature and the visions which nurture all progress.

"In 1492, the discovery of the American Continent transformed geography and the course of human events. Today, conquest of ultraterrestrial space—with its attendant unknowns—recreates our perspectives and enhances our paradigms.

"Mexico, while expressing its hope that this human achievement will result in good for mankind and that all the peoples on Earth will participate in its fulfillment with clear conscience of their common destiny, for the development of this new stage, offers not a power nor a richness it does not possess but the moral heritage decanted from its own history: an unquenchable thirst for material and spiritual improvement and an unyielding faith in the supremacy of reason and justice as a way and an inspiration for human conduct which now has attained a new far reaching responsibility."

From J. J. Trejos Fernandez, president of Costa Rica

"I join in the wish of all Costa Ricans for the success of the historical exploit to be carried out by Apollo 11, in that it represents the scientific and technical progress attained by man in his peaceful struggle for the conquest of space and in that the crew of this ship represents human valor, will, spirit of adventure and ingenuity.

"The enormous scientific and technical effort deployed in order to take the first men to the moon deserves the gratitude of mankind because from this effort will come new benefits for improving the well-being of the human race.

"With faith we hope for better days for all mankind if there is later added to this successful endeavor—new determination for justice and liberty, as they correspond to the respect owed each human being and in favor of a major diffusion of love of one's neighbor, whose efforts we can hope will be stimulated by the spirit of humanity derived from a more clear and vivid awareness of the minuteness of this planet, which serves as our home in the cosmos.

"As representative of the Costa Rican nation, I extend my greetings to the heroes of Apollo 11 and to all those who are making this historical feat possible."

SOUTH AMERICA

From Artur da Costa e Silva, president of Brazil

"In rejoicing together with the government and the people of the United States of America for the event of the century, I pray God that this brilliant achievement of science remain always at the service of peace and of mankind."

From Carlos Lleras Restrepo, president of Colombia

"As you prepare to undertake one of the most extraordinary feats in history, I wish to send to you on behalf of the people and the Government of Colombia, a warm greeting with our wishes for the complete success of your mission. I also want to express the admiration of all Colombians for your personal heroism, for the scientists and technicians that have contributed their knowledge to this enterprise and for the great North American nation whose support has made possible a project that only yesterday appeared to be unfeasible.

"Please leave on the moon along with the other objects that will bear witness of man's first arrival to our satellite, this message, as a symbol of friendship between Colombia and the United States.

"You will descend upon the moon on our national holiday, when we observe the 159th anniversary of our independence. We, in Colombia, will be honoring the memory of the patriots that changed the course of our history on the same day when you will be writing an immortal page in the annals of mankind."

From Linden Forbes Sampson Burnham, prime minister of Guyana

"To those coming after: We cannot tell on what future day—beings of our own kind or perhaps from some other corner of the cosmos, will come upon this message but for those coming after, we wish to record three things:

"First, we salute these astronauts, the first two of our human race who with faith and courage have voyaged far beyond the familiar limits of our earthly home to the Moon. It is certain that their mission ushers in the greatest adventure of life since its primeval beginnings on this planet, Earth.

"Second, as members of our human race thus thrust among the stars, we pledge ourselves to work towards ensuring that the technology which has made it possible and the resources which may be discovered will be used for the benefit of all mankind irrespective of terrestrial divisions of race or creed or levels of development.

"Third and finally, we wish to set down the facts about the people for whom I speak. We are a small nation of some 700,000 souls living on the shoulder of South America in a country some 83,000 square miles in area. Our ancestors came from nearly every corner of the planet Earth and our people today profess a variety of creeds and of ways of living. But in a world in which divisions deepen and where too often one man's hand is set against his brother, we are proud that we have given to our time an example of how out of diversity we have made one people, one nation—with one destiny.

"By working out this destiny, we have developed institutions based on the recognition of the equality of all men, forms of government in which all can participate and a system of justice which protects the weak. With the help of friendly nations, and working together, we are embarking on the challenging task of abolishing disease and poverty from our midst, and of developing our economy so that it can support a worthy level of living for our people. We have, likewise, striven hard to ensure that men everywhere are free to determine their own way of life.

"We do not know what shall be the judgment of history but we would be well pleased if on some later day when this is read, it is said of us that we strove greatly to advance the dignity of all men."

Besides the eighteen nations from six continents whose letters are quoted above, fifty-three additional nations sent goodwill messages that came to be etched on the Apollo 11 silicon disc. In alphabetical order they were, from *Africa*: Dahomey (today known as the Republic of Benin),

Ethiopia, Ghana, Kenya, Lesotho, Madagascar, Mali, Morocco, Senegal, Sierra Leone, South Africa, Swaziland, Togo, Tunisia, Upper Volta (known since 1984 as Burkina Faso), Zambia; from *Asia*: Afghanistan, Republic of China (Taiwan), Iran, Israel, Japan, Laos, Lebanon, Malaysia, Maldives, Pakistan, Philippines, Thailand, Republic of Vietnam (South Vietnam); from *Europe*: Cyprus, Denmark, Estonia, Greece, Iceland, Ireland, Italy, Latvia, Malta, Netherlands, Norway, Romania, Vatican, Yugoslavia (comprising today's states of Serbia, Croatia, Bosnia & Herzegovina, Slovenia, Montenegro, Macedonia, and Kosovo); from *North America*: Dominican Republic, Jamaica, Nicaragua, Panama; from *South America*: Argentina, Chile, Ecuador, Peru, Trinidad and Tobago, Uruguay. Seven more countries sent in goodwill letters to NASA but they arrived too late to be included on the silicon disc. They came from Gabon, in Africa; from Ceylon (Sri Lanka), in Asia; from Finland, West Germany, Poland, and Sweden in Europe; and from Bolivia, in South America.

Interestingly, among the goodwill letters, a handful came from countries that were part of the so-called Eastern Bloc, the group of socialist countries in Central and Eastern Europe that were satellite states of the Soviet Union. Letters were sent to NASA by Poland's ambassador to the United States Jerzy Michalowski; by Romanian president Nicolae Ceauşescu; and by Yugoslavia's president Josip Broz Tito. Two of the USSR's Baltic republics, annexed under protest by the Soviet Union after World War II, also sent letters for the disc. From Estonia came a letter from veteran diplomat Ernst Jaakson, and from Latvia came one from the head of the Latvian diplomatic service, Anatols Dinbergs.

Given the Cold War rivalries of the period, it should come as no surprise that no letters came from the People's Republic of China or the Soviet Union—the latter, of course, America's main rival during the Space Race era. The subject of Soviet interest in Apollo 11 will be covered in a later chapter that includes correspondence between different Soviet citizens and institutions that were sent to Neil Armstrong in the months and years following the first Moon landing. As readers will discover, Neil and his crewmates had many friends behind the Iron Curtain. Neil in particular was greatly admired by thousands in the communist world—not just for his being a distinguished astronaut and the first man on the Moon but also for his personal integrity and respectful, self-effacing character—whether or not they could publicly say so at the time.

• • •

When in the years 2002 to 2005 I was researching and writing *First Man: The Life of Neil A. Armstrong* (Simon and Schuster, first published in November 2005), I had very little access to Armstrong's correspondence. Especially not to his personal correspondence, which was in Neil's own secure possession either in the basement of his home in suburban Cincinnati or in storage spaces he had been renting for many years in nearby Lebanon, Ohio, a rural community where for some twenty years Neil and his family (with his first wife Janet Shearon Armstrong and sons Rick and Mark) had lived on a farm.

Certainly Neil gave me by far more direct access to him and his materials than any other writer or historian before me, including the opportunity to tape-record interviews with him for a total of some fifty-five hours conducted over several months. During this time I did see many of his papers but relied primarily on the multitude of documents relevant to Neil that existed in various NASA, military, and university archives. The fact was, Neil was Neil—a very private man—and he shared with me only a small fraction of the letters, cards, emails, and other correspondence he had received, and himself written, over the course of his life, and those items I did see were always what Neil chose to share with me. I never had free, direct, or unrestricted access to his personal papers. The same was true when I put *First Man* into a second edition shortly following his death in August 2012. So, I always regarded my biography of Neil—as lengthy as it was, over 700 pages—as incomplete, and in some respects not very well informed.

Neil began donating small parts of his collection to his alma mater in 2008, but the entirety of his papers did not come to Purdue until after his death. As soon as she was able to proceed, considering the very deep grief she felt following her husband's death, Carol Held Knight Armstrong, Neil's wife since 1994, did her absolute best to help Purdue archivists carry out Neil's intention for his papers to be delivered to West Lafayette. It turned out to be a very sizable collection, with 350-plus archival boxes stuffed full with Armstrong materials arriving on the Purdue campus. Neil's bequest included technical documents, coursework, NASA working papers and subject files, writings and notes, scripts for speeches, photographs and newspaper clippings, and a wide range of material related not just to Neil's years as an astronaut but to his college education, navy career, training and career as a test pilot, career as a professor of engineering, and later roles as advisor and board member for various industries, businesses, and commissions.

Today the Neil A. Armstrong papers collection is a treasured legacy within the Purdue University Archives and Special Collections' larger Barron Hilton Flight and Space Exploration Archives, established in 2011 with generous support from Mr. Barron Hilton and the Conrad N. Hilton Foundation. The Hilton Archives incorporates not just Armstrong's papers but also the papers of a number of prominent engineers, aviation professionals, scholars, and astronauts, including Purdue graduate Eugene Cernan, the last man to walk on the Moon. It is a historically significant aerospace collection second to none at an American university, one that features a deep fount of primary source materials on the history and development of powered flight. It is also a unique archive in that it dates back to an inaugural gift to Purdue, in 1940, of the papers of pioneering aviator Amelia Earhart, gifted by her husband, George Palmer Putnam. From the Earhart donation, the materials housed at Purdue related to the history of flight grew steadily over the next eighty years as new generations of Purdue faculty and alumni added their contributions to the development of aerospace sciences and associated flight technologies.

For four straight summers from 2015 to 2018, I resided for many days in the Purdue archives, taking a very close look at the Armstrong papers, really fully for the first time. Quickly I became especially fascinated with the approximately 70,000 pieces of fan mail, which Neil had begun to receive in high volume following the Moon landing and that kept coming to him, in bulk, from all around the world, for the rest of his life. Thanks to the good graces of archivists Sammie Morris and Tracy Grimm, I managed to browse all of that correspondence, determined to find the most insightful and fascinating letters written to Neil, as well as a representative sampling of his replies.

The result of my efforts is the book you have in hand, with the letters arranged into chapters according to coherent themes. Most letters are annotated to add context and content to the letters, letter writers, or Neil's response. Some appear as a series of letters, when the exchange of correspondence grew beyond just single letters to Neil and back. Mostly I let the letters speak for themselves and allow readers to draw their own conclusions and arrive at their own insights. But occasionally I insert my own analysis and interpretations upon a piece of correspondence, especially when I believe we need to explore its social and cultural meaning. Certainly it is my own conclusion that the letters ultimately tell us more about ourselves than they do about Neil. The letters illuminate what we wanted from, what we expected from, and what we believed about, fairly or unfairly, our "world iconic hero," Neil Armstrong.

Readers should also be aware that this book is not the last word on Neil Armstrong's letters. The thematic coverage of this book does not come close to covering all the major topics found in the Armstrong correspondence. In the coming months I will work with the Purdue University Press to publish at least one additional book of letters, which will contain thematic chapters focusing on the subjects of religion and belief, conspiracy theories and UFOlogy, correspondence with astronauts, space program officials, and notables from the world of aviation, and Neil's contacts and experiences in the corporate world of business and finance.

This book of letters adds many new insights into the life and times of Neil Armstrong, providing tidbits and informational items. Reading the letters has not changed any of my essential interpretations of Neil's biography, but it has extended and deepened my knowledge of his life in many worthwhile and exciting ways. It has certainly strengthened my feeling for the spot-on appropriateness of the epigraph I used for my book *First Man* with its original publication in 2005. The quote came in the form of a powerful thought from the American scholar Joseph Campbell: "The privilege of a lifetime is being who you are."

Without the generous help and support of Tracy Grimm, associate head of Archives and Special Collections and the Barron Hilton Archivist for Flight and Space Exploration; Sammie Morris, Purdue university archivist and head of Archives and Special Collections; Katherine Purple, editorial, design, and production manager at the Purdue University Press; and Justin Race, the director of the Purdue University Press, this book could not have materialized at all, let alone as beautifully as it did. I also want to thank Kelley Kimm of the Purdue University Press for her outstanding editorial work and patience. My daughter, Jennifer Hansen Gray, once again helped her father in various ways, including not just transcribing many of the letters but helping me to choose the best letters to publish. Rick Armstrong, Neil's oldest son, also helped me a number of times, providing additional information and answering questions prompted by the letters.

Finally, I want to thank all of the people who wrote letters to Neil from 1969 until his death in 2012, and in particular the significant number of folks I was able to hunt down and talk to about their Dear Neil letters.

James R. Hansen
Auburn, Alabama
July 2019

1
FIRST WORDS

At 04:13:24:48 elapsed time into the mission of Apollo 11, which in the United States was a few seconds before 10:57 p.m. EDT on Sunday, July 20, 1969, Neil Armstrong took his historic first step onto the lunar surface. He then spoke his eternally famous first words. What the world heard, a quarter of a million miles through space, was "That's one small step for man, one giant leap for mankind."[1]

No one knew what Armstrong would say when he stepped onto the lunar surface, not even his crewmates. Buzz Aldrin recalled: "On the way to the Moon, Mike and I had asked Neil what he was going to say when he stepped out on the Moon. He replied that he was still thinking it over."[2]

To the end of his life, Armstrong maintained that he did not compose what he would say until sometime after he and Buzz successfully executed the landing of their lunar module, *Eagle*, onto the Sea of Tranquility, which had occurred some six and a half hours earlier, at 4:17:39 p.m. EDT. As Neil explained: "Once on the surface and realizing that the moment was at hand, fortunately I had some hours to think about it after getting there. My own view was that it was a very simplistic statement: what can you say when you step off of something? Well, something about a step. It just sort of evolved during the period that I was doing the procedures of the practice takeoff and the EVA prep and all the other activities that were on our flight schedule at that time. I didn't think it was particularly important, but other people obviously did. Even so, I have never thought that I picked a particularly enlightening statement."[3]

Many people had had ideas for what Armstrong should say when he stepped out onto the Sea of Tranquility—and several of them shared them with him. One to do so who had some official authority over NASA was Willis Shapley, the associate deputy administrator at NASA Headquarters. On April 19, 1969, by which time it was growing increasingly clear that a Moon landing would be attempted in the early summer, Shapley sent a memorandum to Dr. George Mueller, head of the Office of Manned Space Flight. It was a three-page memo entitled "Symbolic Items for the First Lunar Landing," and in it Shapley addressed point-by-point what sorts of items should be left on the Moon by the Apollo Moon landing crew as well as what commemorative articles should be taken to the lunar surface, which ones should be left there, and which ones should be returned. Early in the memo, in discussing what sort of bigger message the Moon landing should present to the world, Shapley wrote: "The intended overall impression of the symbolic activities and of the manner in which they are presented to the world should be to signalize the first lunar landing as an historic step forward for all mankind that has been accomplished by the United States of America. . . . The 'forward step for all mankind' aspect of the landing should be symbolized primarily by a suitable inscription to be left on the Moon and by statements made on Earth . . . and also perhaps by leaving on the Moon miniature flags of all nations." Dr. Mueller passed Shapley's memo on to Deke Slayton, the chief of the astronaut office (and one of the original Mercury astronauts), who shared it with Armstrong.[4] Thus came the conjecture that the seed for the idea for Neil's expression of his "one small step" was planted by the Shapley memo.

But Armstrong never had any recall of the memo—none. As hauntingly similar as the phrase "forward step for all mankind" might seem to be to us, Neil did not remember getting a copy of it or ever hearing anything about it. It seems to be an example—like CBS news anchor Walter Cronkite's comment on television the morning of the Apollo 11 landing about taking "a giant leap"—of a similar statement having been made independently of the thought process behind Armstrong's own words. As Neil would explain over the years: "My guess is that you can take almost any statement, and if you look around for a while, you can find other statements that were made similarly by other people." In Neil's mind, there was never any particular context for the genesis of his phrase and he did not connect it to any other quotation or experience. "Not that I know of or can recall," Neil would say. "But you never know subliminally in your brain where things come from. But it certainly wasn't conscious. When an idea runs for the first time through your own mind, it comes out as an original thought."[5]

Nor was Neil aware of the article "Le Mot Juste for the Moon" ("The Right Word for the Moon") by journalist William H. Honan, culture editor at the *New York Times*, published in the July 1, 1969, issue of *Esquire* magazine. Honan's article began with this extraordinary appeal to Neil Armstrong: "We, the human race, hereby request that the first man on the moon, destined to speak on our behalf, pause for a moment and give some consideration to what he intends to say."[6] Honan then offered to Neil "Fifty Helpful Hints," some serious, some in jest, that he had solicited from diverse notable individuals, including the following:

Hubert H. Humphrey, former vice president of the United States: "May the conflicts and troubles of man never find a home here. May the moon be a symbol of peace and cooperation among the nations of earth."

William Bernbach, chairman of the board of the international advertising agency Doyle Dane Bernbach: "This neighborhood is never going to be the same."

James Whittaker, the first American to climb Mount Everest: "Now how the hell do we get back?"

Sun Ra, space-age jazz poet: "Reality has touched against myth / Humanity can move to achieve the impossible / Because when you've achieved one impossible the others / Come together to be with their brother, the first impossible / Borrowed from the rim of the myth."

Keir Dullea, the actor who played Major Tom in the 1968 Stanley Kubrick film 2001: A Space Odyssey: "I shall never lose the awe of being the first man to stand on a given spot where no man had ever trod before."

Lawrence Ferlinghetti, San Francisco beat poet: "We Roman emperors of space have hereby proved that heaven doesn't exist and that the only god is consciousness itself."

Marianne Moore, American modernist poet: "Just got here and I have to look around."

Leonard Nimoy, actor who played Spock in the television series Star Trek: "I'd say to earth, from here you are a peaceful, beautiful ball and I only wish everyone could see it with that perspective and unity."

E. H. Munn Sr., chairman of the executive committee of the Prohibition National Committee: "Be assured, people of earth, we shall not corrupt the moon with beverage alcohol, with tobacco's poisons or with other of man's unfortunate concoctions. Rather, we shall keep this area of God's universe pure and free from the ugliness and the devastation of the sinful excesses of humankind."

Robert Graves, British poet, historical novelist, critic, and classicist: "Forgive the intrusion, Ma'am" (a tactful propitiation of the Moon Goddess)."

Harold O'Neill, president of the American Sunbathing Association: "I proclaim this a wonderful spot where the bare facts of life shall not be loused up with earthly convention. Moonlife shall be sans clothing, thus eliminating the need for vacation wardrobes . . . an important weight factor when considering space travel."

New York poet **Stanley Kunitz**: "Earth was my home, but even there I was a stranger. This mineral crust. I walk like a swimmer. What titanic bombardments in those old astral wars! I know what I know: I shall never escape from strangeness or complete my journey. Think of me as nostalgic, afraid, exalted. I am your man on the moon, a speck of megalomania, restless for the leap towards island universes pulsing beyond where the constellations set. Infinite space overwhelms the human heart, but in the middle of nowhere life inexorably calls to life. Forward my mail to Mars. What news from the Great Spiral Nebula in Andromeda and the Magellanic Clouds?"

U.S. senator **George McGovern**: "I raise the flag of the United Nations to claim this planet for all mankind and to signal a new era of understanding and cooperation among nations—both on the Moon and on Earth."

William Safire, speechwriter for President Nixon: "Free at last."

Isaac Asimov, American writer of science fiction and popular science: "Goddard, we are here! (A salute to American rocket pioneer Robert Hutchings Goddard, 1882–1945)

Dr. Timothy Leary, American psychologist, anti-war activist, and writer known for advocating the exploration of the therapeutic potential of psychedelic drugs: "The C.I.A. really blew it again. How did all those barefoot, long-haired, smiling-eyed kids get up here ahead of us?"

Heavyweight champion boxer **Muhammad Ali**: "Bring me back a challenger, 'cause I've defeated everyone here on earth."

Entertainer **Bob Hope**: "1) Well, at least I didn't end up in Havana, 2) My God, smog!, 3) I'll be darned, it's made of cheese!"

Brother Antoninus (aka William "Bill" Everson), an American poet of the San Francisco Renaissance: "Bone cold. An immense Golgotha. Out of this tomb, what resurrection? Out of this dust, what weird rebirth?"

Joseph Heller, author of the satiric anti-war novel Catch 22: "I'd like to hear nothing; the chances are I won't be listening. I'm more interested in what Joe Namath or George Sauer has to say about anything, and I hope the moon landing doesn't take place during a Jets football game and interrupt the telecast. One of my favorite statements of recent times, in fact, came from George Sauer. He was talking about a Baltimore player with a crew cut, and he said: 'He ought to let his hair grow, he looks funny that way.' I doubt if anything said about the moon landing will make more sense." (Joe Namath was the star quarterback of the New York Jets and George Sauer Jr. was the wide receiver on 1968 Jets team that upset the Baltimore Colts in the 1968 Super Bowl played on January 12, 1969. Both Namath and Sauer wore their hair stylishly long.)

Entertainer **Tiny Tim** (aka Herbert Buckingham Khaury), an eccentric American singer and ukulele player best known for his song "Tiptoe through the Tulips": "The first thing I would like to hear him say is 'Praise the Lord through Christ that we landed well and safely.' Then I'd like him to describe the moon, and how it looks and feels, what the craters are like, whether there are any cities around, if there's any air to breathe. I think there are definitely beings on the moon. They will probably be very different from us, but the astronauts should be prepared to show them goodwill of the people of Earth. Things like candies, balloons, bubble pipes, soap bars, pens, pencils, plants, even a ukulele, and, most important, the Holy Scriptures, so we can give our new acquaintances some idea of what life is like down here. I really believe life exists on every planet, even the suns, and before we go visiting other worlds, we should be sure we are ready to make the people we find waiting there our friends." (Interestingly, upon the safe return of Apollo 11 to Earth, Armstrong, in the lunar quarantine facility would play a ukulele, though there is no evidence to connect it with anything to do with Tiny Tim. It would be interesting to know if Neil played "Tiptoe through the Tulips" for his crewmates in the LQF, Aldrin and

Collins, as the silly little song [actually composed in 1929] rose to as high as #17 on the popular music charts in 1968.)

Author **Truman Capote**, best known for his 1966 bestselling book In Cold Blood: "If I were the first astronaut on the moon my first remark would be: So far so good."

U.S. Supreme Court Justice **William O. Douglas**: "I pledge that we the people of the earth will not litter, pollute and despoil the moon as we have our own planet."

Theodore Weiss, American poet: "Moon that we have for thousands of years looked up to, now help us to see the earth in its true light, as whole and one."

New York congressman **Ed Koch**: "I proclaim the moon an international scientific laboratory, for all men of all nations to use peacefully in their quest for a deeper understanding of the many worlds which are theirs."

Gwendolyn Brooks, Pulitzer Prize–winning poet: "Here there shall be peace and love."

David Slavitt, American writer, poet, and film critic: "We have realized an ancient dream, and it is rock and dust; now we must look back to earth, imagine what it ought to be, and hope that dream turns out better."

American economist **John Kenneth Galbraith**: "We will hafta pave the damn thing."

Marshall McLuhan, Canadian professor, philosopher, and public intellectual whose work set one of the cornerstones of the study of media theory, known for coining the expression "the medium is the message" and the term "global village": "The thickest mud that was ever heard dumped."

Hollywood actor **Sal Mineo**: "Oh, hi!"

Russian American writer and philosopher **Ayn Rand**: "What hath man wrought?"

Norman Cousins, American political journalist, author, and world peace advocate: "Actually, I would hope he might feel the occasion called for a moment or two of quiet, perhaps even meditation."

Russell Baker, Pulitzer Prize–winning American writer known for his satirical commentary and self-critical prose: "I hereby declare this moon open for pollution."

R. Buckminster Fuller, American architect, systems theorist, inventor, and futurist who coined the term "Spaceship Earth": "Wish you were here."

United Nations secretary general **U Thant**: "I would like the first astronaut to land on the moon to remind us again that we are all brothers, so that we may all have a new sense of perspective to enable us, in the language of the Charter of the United Nations, to practice tolerance and live together in peace with another as good neighbors."

W. H. Auden, Anglo-American poet: "I've never done this before!"

Father Malcolm Boyd, Episcopal priest and author of Are You Running With Me Jesus?: "I baptize thee in the name of the Father, and of the Son and of the Holy Ghost. Amen. All right, you guys, whoever you are standing behind the rock over there, come out with your hands up!"

The correspondence that follows comprises mostly telegrams sent to Neil in the weeks immediately preceding the launch of Apollo 11 on July 16, 1969, many offering suggestions for what he should say when he first steps onto the Moon. A major theme of these suggestions is that he should say something spiritual or even quote scripture. In fact, as a multitude of letters in the Armstrong Collection at Purdue University attest, Neil would receive a great many letters over the years concerning God and the possible religious meaning of landing on the Moon and of space exploration generally. However, he would respond to virtually nothing that came to him concerning religion—any religion. Though his mother, Viola Engel Armstrong (1907–1990), was a devout evangelical Christian (a member of the Reformed Church in America), Neil was a deist, a person whose belief in God was founded on reason rather than on revelation, and on an understanding of God's natural laws rather than on the authority of any particular creed or church doctrine. Similar to many other deists (famous deists of the Enlightenment include Englishman John Locke, Scotsman David Hume, and a number of founding fathers of the United States, notably James Madison, Alexander Hamilton, and Thomas Jefferson), Neil, a highly private man, kept his views on God to himself and did not care to hear the religious views of others.

"EVERYONE HAS BEEN TRYING TO GIVE
YOU THE FIRST WORDS"

June 21, 1969

Dear Col. Armstrong:

I see in Esquire that everyone has been trying to give you the first words to be spoken on the moon. How about these un-Cernan-like suggestions?

"I've taken Man's first step into the Universe, and the ground is solid." (Or squashy, but firm, as the case may be.)

"God has allowed us to take Man's first steps into the Universe. We are humble and grateful."

I dedicate the enclosed recent article of mine to you, and I dare believe you will find these words as succinct, prophetic and perhaps eloquent as any that will be written later about your destined deed.

If you want, I'll frame these two pages for your wall, and deliver it to you."

Sincerely,

Thomas Sweeney
Wheeling, West Virginia

The letter writer seems to have been Thomas B. Sweeney Jr. (1903–1973), a Republican politician from Wheeling, West Virginia, whose grandfather Thomas Sweeney was a prominent early industrialist (glassmaker) in Wheeling and served in both houses of the Virginia General Assembly as a Whig. This Thomas Sweeney was a member of the West Virginia Senate 1st District (1939–1942) and at various times also a candidate for West Virginia House of Delegates, U.S. senator from West Virginia, and U.S. representative from West Virginia. He was also a delegate to the Republican National Convention from West Virginia in 1948 and 1960. Mr. Sweeney's reference to "Cernan" is to astronaut Eugene Cernan, who, prior to the date of this letter, had served as pilot for Gemini 9 (June 1966) and as lunar module pilot for Apollo 10 (May 1969). Mr. Sweeney's mention of "un-Cernan-like suggestions" refers to a colorfully worded comment by Cernan during an unexpected tumble in the Apollo 10 lunar module during a maneuver in preparation for rendezvous and docking with the command module. While still on a hot mike, with the world listening, Cernan blurted out, "Son of bitch, what the hell happened?"

"CARRY A SUPPLY OF ROCKET POPCORN SEED"

Western Union Telegram
To: "Commander, Neil A. Armstrong, Astronaut"
Received NASA Kennedy Space Center Communication Center
July 1, 1969, 8:14 A.M. CDT
Sent from Murray, Kentucky

MR. ARMSTRONG, WE RESPECTFULLY REQUEST THAT APOLLO 11 CARRY A SUPPLY OF ROCKET POPCORN SEED FOR PLANTING ON THE MOON. ROCKET POPCORN HAS BEEN ON THE MARKET FOR OVER 20 YEARS AS WE LONG AGO ANTICIPATED THE MOON LANDIN. WE MUST BE FIRST ON THE MOON WITH THE ALL-AMERICAN PRODUCT POPCORN WHICH HAS TREMENDOUS FOOD ENERGY VALUE AND WILL GLADLY SUPPLY THE ROCKET POPCORN SEED FOR PLANTING. ALSO A YEARS SUPPLY OF ROCKET POPCORN FOR POPPING FOR ANY ASTRONAUT WHO DECIDES TO STAY ON THE MOON. WE HAVE WIRED PRESIDENT NIXON FOR HIS APPROVAL AND YOUR CONSIDERATION WILL BE GREATLY APPRECIATED.

MOST RESPECTFULLY
ED CHRISMAN
CHRISMAN POPCORN CO
MURRAY KY

"ANNOUNCE ALL UNITS IN THE INTERNATIONAL SYSTEM (METRIC)"

Western Union Telegram
To: "Neil A. Armstrong, Commander, Apollo 11"
Received NASA Kennedy Space Center Communication Center
July 1, 1969, 2:40 P.M. CDT
Sent from Hattiesburg, Mississippi

TWO THOUSAND STUDENTS AT THE UNIVERSITY OF
SOUTHERN MISSISSIPPI PLEAD WITH YOU TO ANNOUNCE
ALL UNITS IN THE INTERNATIONAL SYSTEM (METRIC)

JOHN M FLOWERS
ASSOCIATE PROFESSOR
UNIVERSITY OF SOUTHERN MISSISSIPPI
HATTIESBURG MS

"I *BEG* YOU NOT TO GO"

July 8, 1969

Astronaut Neil Armstrong
c/o NASA
Houston, Texas 77001

Dear Commander Armstrong:

I am writing this letter hoping you will give it more consideration than
would be expected. I am a 12 year old girl living in a small town of
8,200. I know you must be a brilliant man and your intelligence and
position leaves me feeling quite insignificant. However, after considering
all this I am compelled to write to you anyway.

The night of June 11, 1969 I had a very disturbing dream. It had
such an impact upon me every time I start to tell it I cannot hold back

the tears. The dream was so forceful it woke me out of my sleep. Then I prayed and asked God to show me if it was Him. He then proved to me it was by making his picture glow in the dark and giving me another dream which came true. I am a born again Christian and have experienced God's transforming power in my life. This is why I am willing to do what He tells me and knew He is concerned about you and Aldrin and Collins.

In the dream I was told to tell you not to go on the flight. God showed me it would have an effect upon the whole world. Even though I don't know you at all, I *beg* you not to go. Mr. Armstrong, *please* don't go for your sake, the world, and the other two.

If you will give the enclosed copies of this letter to Aldrin and Collins, I would be very grateful. Thank you.

Praying for you,

Georgia Teal
c/o Vealton Teal
Burley, Idaho

"THANK GOD PUBLICALLY"

Western Union Telegram
To: "Astronaut Neil Armstrong"
Received NASA Kennedy Space Center Communication Center
July 9, 1969, 1:39 P.M. EDT
Sent from Cincinnati, Ohio

PLEASE ACKNOWLEDGE AND THANK GOD PUBLICALLY
WHEN YOU LAND ON THE MOON

MRS JAMES STOUT
CINCINNATI OHIO

"ALWAYS KNEW THERE WAS A MAN ON THE MOON"

Western Union Telegram
To: "Astronaut Neil Armstrong"
Received NASA Kennedy Space Center Communication Center
July 11, 1969, 9:11 A.M. EDT
Sent from Lancaster, California

ALWAYS KNEW THERE WAS A MAN ON THE MOON. UP TO
NOW DIDNT KNOW HIS NAME. AT LEAST YOUR SECRETARY
CAN KEEP TRACK OF YOU WHICH WAS MORE THAN YOUR
FORMER GAL FRIDAY COULD DO. HAPPY LANDING NEIL
DEAR AND SAFE RETURN. SHALL BE WATCHING AND
LISTENING EVERY MINUTE.

DELLA MAE BOWLING

"SAY A PRAYER FOR HUMANITY"

Western Union Telegram
To: "Apollo 11 Mission Commander Neil Armstrong"
Received NASA Kennedy Space Center Communication Center
July 11, 1969, 5:58 P.M. EDT
Sent from San Francisco, California

ALMIGHTY GOD BLESS YOU AND THE CREW ON YOUR
JOURNEY TO THE MOON[.] WHEN ON THE MOON PLEASE
SAY A PRAYER FOR HUMANITY

JOHN SHANOVSKOY
ARCHBISHOP OF SAN FRANCISCO AND WESTERN UNITED
STATES RUSSIAN ORTHODOX GREEK CATHOLIC CHURCH OF
AMERICA

"PLEASE TAKE MEMENTO OF JOHN F KENNEDY"

Western Union Telegram
To: "Astronaut Neil Armstrong"
Received NASA Kennedy Space Center Communications Center
July 11, 1969, 9:11 04 P.M. EDT
Sent from Toledo, Ohio

PLEASE TAKE MEMENTO OF JOHN F KENNEDY TO MOON[.]
HE GAVE INITIATIVE TO GO GO GO

MARY KATHY AND ROBERTA STOCKWELL
FELLOW OHIOANS
TOLEDO OH

"WHAT GOD HATH WROUGHT"

Western Union Telegram
To: "Neil Armstrong"
Received NASA Kennedy Space Center Communication Center
July 15, 1969, 7:51 A.M. EDT
Sent from Jackson Heights, New York

SUGGEST FOR WHAT TO SAY WHAT GOD HATH WROUGHT

HARRISON A MOYER
PRESIDENT OF THE SAMARITANS
JACKSON HEIGHTS NY

"KNEEL AND GIVE HIM THANKS"

Western Union Telegram
To: "Astronaut Neil Armstrong"
Received NASA Kennedy Space Center Communication Center

July 15, 1969, 8:48 A.M. EDT
Sent from North Bergen, New Jersey

THE WHOLE WORLD AND GOD WILL BE WATCHING YOU.
PLEASE PLEASE AS YOUR FIRST ACT KNEEL AND GIVE HIM
THANKS

MR EDWARD SOMICK
NORTH BERGEN NJ

"IT IS THE PILOT'S FIRST DUTY AFTER LANDING"

Western Union Telegram
To: "Neal Armstrong"
Received NASA Kennedy Space Center Communication Center
July 15, 1969, 8:00 A.M. CDT
Sent from San Antonio, Texas

AS A REMINDER UNDER RULE 4.10 IT IS THE PILOTS FIRST
DUTY AFTER LANDING AND SECURING HIS AIRCRAFT TO
NOTIFY CONTEST HEADQUARTERS GIVING PERTINENT
LANDING DATA[.] UNDER RULE 4.11 THE LOCATION OF
A LANDING AWAY FROM THE CONTEST SIGHT MUST BE
CONFIRMED BY 2 IMPARTIAL WITNESSES[.] IT IS EXPECTED
YOU WILL COMPLY WITH THESE REQUIREMENTS. IF
SCORING OF YOUR FLIGHT IS TO BE ACCEPTABLE TO SSA
USE YOUR CONTEST NUMBER. WE HOPE YOU FIND EACH
THERMAL K AT THE DESIRED AND REQUIRED TIME AND
PLACE[.] BEST OF LUCK

H. MARSHALL CLAYBOURN
COMPETITIONS DIRECTOR
SOARING SOCIETY OF AMERICA (SSA)

"REQUEST READING PSLAMS"

Western Union Telegram
To: "Neil Armstrong"
Received NASA Kennedy Space Center Communication Center
July 15, 1969, 9:05 EDT
Sent from Mysore, State of Karnataka, India

REQUEST READING PSLAMS EIGHT CHAPTER 5 VERSES
THREE TO SIX FROM MOON

BOWRON BIBLE CLASS
YADAGIRI
MYSORE
STATE OF KARNATAKA INDIA

"GENESIS IST CHAPTER IST VERSE"

Western Union Telegram
To: "Astronaut Neil Armstrong, Commander of Apollo 11"
Received NASA Kennedy Space Center Communication Center
July 15, 1969, 5:44 P.M. CDT
Sent from Grand Bay, Alabama

MAY I SUGGST THAT WHEY YOU FIRST SET FOOT ON
THE MOON TO GIVE GOD CREDIT FOR THE SUCCESSFUL
ADVENTURE READING FROM THE BIBLE GENESIS 1ST
CHAPTER 1ST VERSE AND ALSO 28TH VERSE[.] THANKS

DEWITT DEES
GRAND BAY ALA

"BRING ALL MEN TOGETHER IN PEACE AND FRIENDSHIP"

Western Union Telegram
To: "Neil Armstrong"
Received NASA Kennedy Space Center Communication Center
July 16, 1969, 6:58 A.M. EDT
Sent from Bowling Green, Ohio

THE LISTENERS OF WSBN RADIO CENTRAL PENN SUGGEST
THESE WORDS BE THE FIRST FROM THE MOON QUOTE LET
US PROCLAIM THAT WE PEOPLE FROM THE PLANET EARTH
SHALL USE THIS MOON AS A LAUNCHING BASE TO BRING
ALL MEN TOGETHER IN PEACE AND FRIENDSHIP UNQUOTE
BEST OF LUCK

WSBN RADIO
BOWLING GREEN OH

"I CAME I SAW I CONQUERED"

Western Union Telegram
To: "Neil Armstrong"
Received NASA Kennedy Space Center Communication Center
July 16, 1969, 7:15 A.M. EDT
Sent from West Palm Beach, Florida

IF YOU ARE WORRIED ABOUT WHAT TO SAY UPON LANDING
OF THE MOON ITS SIMPLE. I CAME I SAW I CONQUERED.
BEST OF LUCK TO ALL OF YOU

NANCY PARSONS
WEST PALM BEACH FL

"IT IS NOW YOUR MOMENT"

U.S. Government Telegram
To: "Commander Neil Armstrong and Apollo 11 Astronauts"
Received NASA Kennedy Space Center Communication Center
July 16, 1969, 8:17 A.M. EDT
Sent from the White House, Washington, D.C.

ON THE EVE OF YOUR EPIC MISSION, I WANT YOU TO
KNOW THAT MY HOPES AND MY PRAYERS—AND THOSE OF
ALL AMERICANS—GO WITH YOU. YEARS OF STUDY AND
PLANNING AND EXPERIMENT AND HARD WORK ON THE
PART OF THOUSANDS HAVE LED TO THIS UNIQUE MOMENT
IN THE STORY OF MANKIND; IT IS NOW YOUR MOMENT
AND FROM THE DEPTHS OF YOUR MINDS AND HEARTS AND
SPIRITS WILL COME THE TRIUMPH ALL MEN WILL SHARE.
I LOOK FORWARD TO GREETING YOU ON YOUR RETURN.
UNTIL THEN, KNOW THAT ALL THAT IS BEST IN THE SPIRIT
OF MANKIND WILL BE WITH YOU DURING YOUR MISSION
AND WHEN YOU RETURN TO EARTH.

RICHARD M. NIXON
PRESIDENT OF THE UNITED STATES

"WISHING YOUR COMPLETE SUCCESS"

U.S. Government Telegram
To: "Mr. Neil A. Armstrong"
Received NASA Kennedy Space Center Communication Center
July 16, 1969, 8:49 A.M. EDT
Sent from the Pentagon, Washington, D.C.

PLEASE PASS THE FOLLOWING MESSAGE TO MR. NEIL A.
ARMSTRONG IMMEDIATELY PRIOR TO APOLLO 11 LAUNCH.

PERSONAL FOR MR. ARMSTRONG FROM MCCONNELL

SECRETARY SEAMANS AND I SPEAK FOR ALL MEMBERS
OF THE UNITED STATES AIR FORCE IN WISHING YOUR
COMPLETE SUCCESS THROUGHOUT YOUR HISTORY-
MAKING MISSION TO THE MOON AND RETURN. YOUR
PROGRESS THROUGH ALL THE PHASES OF ARDUOUS
PREPARATION FOR THIS FLIGHT IS IN ITSELF A NOTABLE
AND PRAISEWORTHY ACHIEVEMENT. IN UNDERTAKING
MAN'S FIRST LANDING UPON THE LUNAR SURFACE, YOU
WILL EARN STILL FURTHER GRATITUDE AND HONOR FROM
OUR NATION AND THE WORLD. GODSPEED.

J.P. MCCONNELL, GENERAL, USAF
CHIEF OF STAFF

"THIS IS A FIRM CONTRACT"

Western Union Telegram
To: "Neil Armstrong"
Received NASA Kennedy Space Center Communication Center
July 18, 1969, 7:54 A.M. EDT
Sent from Las Vegas, Nevada

THIS IS A FIRM CONTRACT OFFER[.] I AM OFFERING 120,000
EACH FOR APPEARANCES AT THE ED SULLIVAN SHOW AND
HOLLYWOOD PALACE[.] EACH APPEARANCE WILL LAST TWO
MINUTES OR IF YOU DESIRE WE CAN EXTEND THE TIME
AS LONG AS YOU WANT[.] I AM THE TALENT SCOUT FOR
THE AERIAL ATTRACTIONS FOR THE ED SULLIVAN SHOW
ORIGINATING IN NEW YORK AND HOLLYWOOD PALACE
HOLLYWOOD CALIF

GOOD LUCK

ALEXANDER DOBRITCH
EXECUTIVE PRODUCER
CIRCUS CIRCUS

2
CONGRATULATIONS AND WELCOME HOME

After eight days in space—about twenty-one and a half hours of which were spent by Armstrong and Aldrin on the lunar surface—the crew of Apollo 11 returned safely to Earth, snug inside their command module. They splashed down into the Pacific Ocean on Thursday, July 24, 1969, at 11:51 Houston time, some 940 nautical miles southwest of Honolulu and 230 miles off Johnston Atoll. Only 13 miles away from the splashdown was the USS *Hornet*, the aircraft carrier designated for the Apollo 11 recovery. Aboard ship was President Richard Nixon along with Secretary of State William P. Rogers and NASA Administrator Thomas O. Paine, who were accompanying the president on a twelve-day around-the-world trip that included a stop in war-torn Vietnam. Also aboard was Admiral John S. "Jack" McCain, commander-in-chief of the Pacific Command (CINCPAC) and the commander of all U.S. forces in the Vietnam theater. For two years, since October 1967, Admiral McCain's son, naval aviator John S. McCain III, the future U.S. senator from Arizona, had been in a North Vietnamese prison camp, where he remained until his release in March 1973.

Waiting for Apollo 11 to hit the water—and "hit" was what it did, like a ton of bricks, forcing a grunt out of each astronaut—were four navy helicopters sent out from the *Hornet*, with a team of three frogmen poised to jump into the sea and bring the astronauts quickly and safely out. The swimmers, who just days before had been working navy demolitions in Vietnam, attached the orange flotation collar around the bobbing spacecraft, then opened the hatch; It was 12:20 p.m. CDT (6:20 in the morning Hawaii time). Armstrong, Aldrin, and Collins felt like they had been in the water for eternity, willing themselves not to be seasick, only twenty-nine minutes having elapsed since splashdown. Into the command module the head of the water rescue team, twenty-five-year-old Lieutenant Clancy Hatleberg, threw biological containment garments, or BIGs. Grayish-green in color, they were the rubberized, zippered, hooded, and visored containment suits meant to save the world from "Moon germs." Each frogman himself wore a special BIG topped by a side-filtered face mask. Swimming in the garment was easy compared to the astronauts' having to put them on inside *Columbia*. Dealing with gravity for the first time in eight days, they were so light-headed, their feet and legs so swollen that they could barely stand, especially against eighteen-knot winds causing heavy waves.

BIGs donned, the Apollo 11 crew squeezed through the small hatch, Armstrong coming out last. Before escorting them one by one into the raft bobbing alongside, the frogmen sprayed them with Betadine. Inside the dinghy the astronauts were given cloths and two different doses of chemical detergent to continue the scrub-down. When they were finished, the frogmen tied the cloths to weights and dropped them into the ocean. Virtually nothing was said by the astronauts because the visors and headgear of their BIGs made it almost impossible to be heard, especially with four helicopters beating their rotors overhead. They sat for fifteen minutes, until a helicopter got the order to pick them up. The *Hornet* was now in view, less than a quarter of a mile away. With television cameras on board a couple of the helicopters, every moment of the recovery was broadcast live around the world.

At 12:57 p.m. CDT, the helicopter landed on the *Hornet*'s flight deck. A brass band played; sailors crowded on deck cheered. A grinning President Nixon stood on the bridge along with Secretary of State Rogers and NASA Administrator Paine.

Still inside the chopper, the astronauts rode one of the ship's elevators down to the hangar deck, then walked down a newly painted line through a cheering crowd of seamen and VIPs into the mobile quarantine facility. The MQF was a thirty-five-foot-long modified house trailer in which they would remain until they arrived at the Lunar Receiving Laboratory (LRL) in Houston on July 27. Inside the MQF they sat in easy chairs where they underwent microbiology sampling and a preliminary medical exam.

There was time only for a quick shower before seeing the president. Following the playing of the national anthem, President Nixon, nearly dancing a jig, addressed the astronauts via intercom. Crouching behind a picture window at the back end of the trailer, the three tired but exhilarated crew members arranged themselves: Neil to the president's left, Buzz to the right, and Mike in the middle. Nixon welcomed them back to Earth and told them that he had called each of their wives the day before to congratulate them. He also invited the astronauts and their wives to a state dinner in Los Angeles. The president closed his remarks by calling the eight days of Apollo 11 "the greatest week in the history of the world since the Creation."[7]

As the ship steamed toward Hawaii, the astronauts could not yet fully relax as there were more postflight medical exams to undergo. A doctor spotted that Neil had an accumulation of fluid in one of his ears. Caused by the stress of reentry, it cleared up by the next day. With the doctors interested in how eight days in zero gravity had affected their bodies, it almost seemed to the astronauts that the mission was still ongoing.

Inside the small living room of the MQF, the astronauts enjoyed a cocktail hour, with Neil drinking scotch. Dinner of grilled steaks and baked potatoes followed. Tired to the bone, they slept snugly that night, in real beds with blissful pillows, for nearly nine hours, arising in the morning with their regular sleeping pattern restored, but only temporarily, as they still had many time zones to cross before making it back home to Houston.

The astronauts stayed two nights on the *Hornet*, Neil enjoying it more than Mike and Buzz as he had come to love life on the high seas while in the navy. Neil played a marathon game of gin rummy with Mike, while Buzz read or played solitaire. For an hour at a time, done repeatedly, they autographed photographs earmarked for NASA and White House VIPs. They also perused some of the many cards, letters, and telegrams that had been mailed to them for receipt on the *Hornet*. A few of those items are included in the contents of this chapter to follow.

The scene at Pearl Harbor was wild as the astronauts arrived on Saturday morning, July 26. Cheering crowds lined the streets, while a brass band played and American flags waved. Atop the *Hornet's* mast was a broomstick, the symbol in the navy of a mission well done. A young boy on crutches, likely a polio victim, scrambled alongside the truck for as far as he could go. At Hickam Air Force Base, a C-141 transport was fueled and waiting for the MQF to arrive and be loaded carefully inside the plane's cavernous belly for the long flight to Texas.

It was closing in on midnight when the C-141 landed at Ellington Air Force Base in Houston. The off-loading of the MQF at Pearl and its loading at Hickam had both gone well, but at Ellington it took three lengthy tries, lasting some ninety minutes, before the trailer containing the astronauts was loaded safely onto the truck. Around 1:30 a.m. the truck finally rolled out of Ellington, slowly heading down NASA Road 1 to the Manned Spacecraft Center. Up and down the road, despite the late hour, throngs of people were still cheering the returning heroes.

It took another hour, until 2:30 a.m., for Neil, Mike, and Buzz to arrive at their destination, the LRL. For the next twenty-one days, until their quarantine was lifted, the crew of Apollo 11 continued their mission—mostly in the form of debriefings—inside the LRL.

The LRL provided a safe, secure, and quiet environment. It had private bedrooms for each crew member, a kitchen, a dining area, a large living room, and a recreation area with a television and a big screen onto which recent Hollywood movies could be projected. With them inside were two cooks, a NASA public relations officer, a medical lab specialist, and a janitor. The space was big enough to accommodate everyone without crowding the astronauts. On August 5, the LRL chef surprised Neil with a cake on his thirty-ninth birthday.

Near the end of their stay, each astronaut, as a federal government employee, filled out an expense report for the flight to the Moon and back. The forms read: "From Houston, Tex., to Cape Kennedy, Fla., to the Moon, to the Pacific Ocean to Hawaii and return to Houston, Tex." The astronauts had traveled by "Government Aircraft, Government Spacecraft, USN *Hornet*, USAF Plane." Their reimbursement was $33.31 each.

While in quarantine, the crew considered really for the first time how all the glamour and publicity would affect their lives and the lives of their families. Just prior to reentry, astronaut Jim Lovell had warned: "Backup crew is still standing by. I just want to remind you that the most difficult part of your mission is going to be after recovery."[8]

They understood Lovell's message. Armstrong would later think back on what he thought at the time: "We were not naïve, but we could never have guessed what the volume and intensity of public interest would turn out to be. It certainly was going to be more than anything any of us had experienced before in previous activities of flight. And it was."[9]

Their quarantine came to an end at 9:00 on Sunday evening, August 10. Going back to the restrictions the crew had been placed under days prior to the launch, they had been in physical seclusion for over a month.

Outside the LRL, a NASA staff car and driver waited to drive them home individually.

Their short trips home that night presaged the astronauts' lot for years to come, especially Neil's. The moment each car passed through the NASA gate, a different television crew pulled behind to follow the famous passengers. Reporters and photographers awaited them in front of their residences.

When they returned to their offices at the Manned Spacecraft Center a few days later, they saw the many stacks of mail that had been delivered for each of them, with Neil's stack by far the biggest. For the next several weeks, Neil would get some 10,000 pieces of mail per day, almost none of it he would be able to answer for months to come. First, from late September to early November, there would be the forty-five-day "Giant Step" tour to twenty-three countries. Then, in December and January, Neil would make a three-week Bob Hope USO tour to entertain the troops in Southeast Asia. Finally, in February 1970, he would make a ten-day trip to the Soviet Union, only the second American astronaut to make an official visit.

Waiting for him back in Houston when he returned from all the traveling were some 300,000 cards and letters that he was now supposed to answer, with tens of thousands more to come.

What follows in this chapter is just the tip of that iceberg: a sampling of the many hundred telegrams that Neil Armstrong and his two crewmates received following their successful return from the first Moon landing. In this sample the reader will encounter telegrams from politicians, military officers, world leaders, space program officials, and other notable individuals. Because they are noteworthy—but may not still be well known to today's readers—I have provided a note immediately following the communiqués respective to these individuals and their organizations. Some who sent telegrams were Neil's personal friends, while others were everyday people who wanted to send their congratulations or make contact with him. Some of the early telegrams in this chapter arrived for him by radiogram aboard the *Hornet*. Others came into the communications center at Kennedy Space Center in Florida or into the central mail room at the Manned Spacecraft Center in Houston. As the reader will see, not all of the telegrams came specifically for Neil; some were addressed to the Apollo 11 crew. But many came expressly for Armstrong. As famous as Aldrin and Collins deservedly became in the public eye, it was Neil who became the global icon.

"I WAS WITH YOU 1000 PERCENT"

Western Union Telegram
To: "Neal F. Armstrong"
Received Central Mail Room, Manned Spacecraft Center, NASA,
Houston, TX
July 23, 1969, 12:26 P.M. CDT
Sent from Junction City, Kansas

CONGRATULATIONS TO THE FIRST MAN TO SET FOOT ON
THE MOON. I WAS WITH YOU 1000 PER CENT. FROM A GOOD
FRIEND OF PRESIDENT RICHARD P. NIXON. I WILL PAY
DEARLY FOR A SAMPLE OF MOON DUST.

RALPH V SJOHOLM 1000 PER CENT AMERICAN &
REPUBLICAN

"FROM . . . THE HIGHLANDS OF VIETNAM"

U.S. Army Radiogram
To: "Armstrong, Aldrin, and Collins"
Received Central Mail Room, Manned Spacecraft Center, NASA,
Houston, TX
July 23, 1969, 2:13 P.M. CDT
Sent from the Highlands of Vietnam

MOST HEARTY CONGRATULATIONS FROM THE OFFICERS
AND MEN OF THE FAMOUS FIGHTING FOURTH DIVISION
OPERATING IN THE HIGHLANDS OF VIETNAM. WE HAVE
FOLLOWED YOUR EPIC FEAT OF LANDING ON THE MOON
WITH IMMENSE PRIDE, AND WITH THE KNOWLEDGE THAT
THIS COURAGEOUS ACCOMPLISHMENT COMPLETES THE
FIRST STEP OF A NEW ERA OF SPACE EXPLORATION; AN ERA
WHICH HOPEFULLY WILL UNITE MANKIND IN THE PURSUIT
OF MUTUAL COOPERATION AND PEACE. WE WISH YOU GOD
SPEED AND A SAFE RETURN TO EARTH.

MG DONN R. PEPKE, CG, 4TH INF DIV

In July 1960, Major General Donn R. Pepke was in command of the U.S. Army's 4th Infantry Division in Vietnam. Born in Minot, North Dakota, in 1917, he served in World War II as a second lieutenant, became chief of staff of the 1st Armored Division in 1962, and commanding general of the 4th Infantry Division in 1968. From 1973 to 1975, at the rank of lieutenant general, he became deputy commanding general of the army forces. The recipient of the Distinguished Service Medal, Silver Star, Legion of Merit, and Bronze Star, General Pepke died in 1995 and was buried in Arlington National Cemetery.

"ALL OF US AT ABC PICTURES"

Western Union Telegram
To: "Astronauts Armstrong, Col. Edwin E. Aldrin, and Lt. Col. Michael Collins"
Received Central Mail Room, Manned Spacecraft Center, NASA, Houston, TX
July 23, 1969, 5:07 P.M. CDT
Sent from Beverly Hills, California

CONGRATULATIONS ON THE MOST MAGNIFICENT ACHIEVEMENT OF MODERN MAN. ALL OF US AT ABC PICTURES CORP WHO ALONG WITH THE REST OF THE RORLD, ARE INDICATED TO YOU FOR YOUR HEROSIM AND SKILL, WOULD BE DELIGHTED AND HONORED TO PROVIDE ENTERTAINMENT DURING YOUR PERIOD OF ISOLATION, IF PERMISSABLE, BY SUPPLYING YOU WITH SIXTEEN MM OR 35 MM PRINTS OF ANY OR ALL OF OUR CURRENT MOTION PICTURES RELEASES[.] MOST SINCERELY

MARTIN BAUM
PRESIDENT ABC PICTURES CORP
1901 AVE OF THE STARS
LOS ANGELES CA

"CULMINATION OF A GREAT NATIONAL EFFORT"

U.S. Air Force Radiogram
To: "Neil Armstrong"
Received CVS-12 USS *Hornet*
July 24, 1969, 9:06 A.M. EDT
Sent from the Pentagon, Arlington, Virginia

PLEASE PASS THE FOLLOWING MESSAGE TO MR. NEIL ARM-
STRONG IMMEDIATELY UPON ON-BOARD RECOVERY.
PERSONAL FOR MR. ARMSTRONG FROM SECRETARY
SEAMANS.
I JOIN AIRMEN AROUND THE WORLD IN CONGRATULATING
YOU UPON ACHIEVING MAN'S FIRST STEPS ON THE MOON.
YOUR MANUALLY CONTROLLED LANDING ON THE LUNAR
SURFACE MARKS THE CULMINATION OF A GREAT NATIONAL
EFFORT. SUPPORTED AND SUSTAINED BY THOUSANDS OF
CIVILIAN AND MILITARY SCIENTISTS AND TECHNICIANS IN
THIS COUNTRY AND ABROAD, YOU HAVE SUCCESSFULLY
COMPLETED THE MOST EXCITING ADVENTURE OF OUR
AGE. YOUR SAFE RETURN TO EARTH USHERS IN A NEW ERA
OF PROMISE FOR ALL MANKING.

ROBERT C. SEAMANS, JR.
SECRETARY OF THE AIR FORCE

*Before becoming the secretary of the air force in 1969, Robert Seamans (1918–
2008) served as NASA's associate administrator from 1960 to 1965 and then
as deputy administrator. Working under NASA Administrator James Webb,
Seamans sometimes also served as the acting administrator. Thus, he was a critical leader during the time of the Apollo program's development. A native of
Massachusetts, Seamans earned a Master of Science degree and a Doctor of
Science, both from MIT. While at NASA he worked closely with the Department
of Defense in research and engineering programs, thereby keeping NASA aware
of DoD military developments and technical needs. He left NASA in 1968
to become a professor at MIT, while remaining a consultant to the NASA
Administrator. After leaving his post with the air force, Seamans became president
of the National Academy of Engineering (May 1973 to December 1974) and the*

first administrator of the new Energy Research and Development Administration, before returning to MIT in 1977 to become dean of the School of Engineering. In 1981 he was elected chair of the board of trustees of Aerospace Corp.

"ALL OHIOANS ARE EXTREMELY PROUD"

Western Union/U.S. Navy Radiogram
To: "Neil Armstrong"
Received CVS-12 USS *Hornet* via San Francisco
July 24, 1969, 9:25 A.M. EDT
Sent from Columbus, Ohio

CONGRATULATIONS TO YOU AND FELLOW ASTRONAUTS ON YOUR MAGNIFICENT JOURNEY TO AND FROM THE MOON. THE CONTRIBUTIONS YOU HAVE MADE TO THE BETTERMENT OF MANKIND ARE JUST BEGINNING TO BE REALIZED AND WILL BE IMMEASURABLE.

ALL OHIOANS ARE EXTREMELY PROUD OF YOUR GREAT ACHIEVEMENT WHICH WILL LIVE IN HISTORY FOREVER.

MOST OF ALL, HOWEVER, WE JOIN WITH YOUR MOTHER AND DAD IN THANKING GOD FOR BRING ALL YOU BACK SAFELY TO EARTH. WE LOOK FORWARD TO THE DAY WHEN YOU COME HOME TO OHIO

JAMES A RHODES
GOVERNOR OF OHIO

For Republican James A. Rhodes (1909–2001), the highlight of his sixteen years as governor of Ohio (1963–1971 and 1975–1983), may very well have been his association with native son Neil Armstrong, who had grown up in various small towns in the Buckeye State in the 1930s and 1940s, notably St. Marys, Upper Sandusky, and Wapakoneta. On September 6, 1969, less than a month after Armstrong and his mates got out of quarantine in Houston, Governor Rhodes played a prominent role in a huge parade in Neil's honor held in Wapakoneta, where the crowd was estimated to be 70,000, more than ten times the town's population. Moreover, even before Apollo 11 had launched, Rhodes had worked with the state legislature to earmark a million dollars for the development of the Neil Armstrong Air and Space Museum in Wapakoneta. The grand opening of the museum, over which Rhodes presided as his pride and joy, was held in July 1972 in conjunction with the third anniversary of Apollo 11.

"ON BEHALF OF THE DEPARTMENT OF DEFENSE"

U.S. Government Radiogram
To: "The Men of the Apollo 11"
Received CVS-12 USS *Hornet*
July 24, 1969, 11:37 A.M. EDT
Sent from the Pentagon, Arlington, Virginia

TO YOU GALLANT AMERICANS WHO JOURNEYED TO THE MOON AND RETURNED IN PEACE, I SEND YOU A HEARTY "WELL DONE" ON BEHALF OF THE DEPARTMENT OF DEFENSE.

I AM SURE THAT ADMIRAL MCCAIN, WHO IS ABOARD THE HORNET, WOULD WANT TO GIVE EACH OF YOU ONE OF HIS CIGARS, BUT THAT MUST WAIT.

WELCOME HOME.

SECRETARY OF DEFENSE MELVIN R. LAIRD

From 1953 to 1969, before he was secretary of defense for President Richard Nixon, Melvin Laird (1922–2016) was a congressman from Wisconsin. As secretary of defense he played a major role in the Nixon Administration's policy for withdrawing U.S. troops from the Vietnam War.

"THE MOST MEMORABLE IN THE HISTORY OF MANKIND"

Western Union International Telegram
To: "Astronaut Neil Armstrong"
Received Central Mail Room, Manned Spacecraft Center, NASA,
Houston, TX
July 24, 1969, 2:31 P.M. CDT
Sent from Reykjavik, Iceland

REMEMBERING OUR MEETING IN THE NEIGHBOURHOOD
OF MOUNT ASKJA I SED YOU MY SINCERE
CONGRATULATIONS ON THE SUCCESS OF OUR VOYAGE
WHICH IS THE MOST MEMORABLE IN THE HISTORY OF
MANKIND AND I JOIN IN THE GENERAL ADMIRATION FOR
THE OUTSTANDING ABILIY AND COUFAGE OF YOU AND
YOUR COLLEAGUES TO WHOM I ASK YOU TO BRING MY
BEST WISHES

BJARNI BENEDIKUTSSON
PRIME MINISTER OF ICELAND

*One of Armstrong's most memorable trips in his training for Apollo 11 was to
the volcanically active and very remote region of central Askja, Iceland. Known
for its volcanic craters (known as "calderas"), the Askja region had a very rocky
terrain with black volcanic sand, as well as a large lake and hot springs. It was
a misty, surreal place unlike anything Neil or the other astronauts who were
with him had ever seen in their travels. Armstrong arrived in Iceland on July 2,
1967, for a geology field trip with twenty-two other astronauts, including Bill
Anders (Apollo 8), Charlie Duke (Apollo 16), Ron Evans (Apollo 17), Fred Haise
(Apollo 13), Ken Mattingly (Apollo 16), Ed Mitchell (Apollo 14), Stuart Roosa
(Apollo 14), Harrison "Jack "Schmitt (Apollo 17), and Al Worden (Apollo 15).*

*Bjarni Benedikutsson (1908–1970) served as prime minister of Iceland from
1963 until his death in July 1970.*

"AS ONE AMERICAN TO ANOTHER"

Western Union Telegram
To: "Astronaut Neil Armstrong"
Received Central Mail Room, NASA Manned Spacecraft Center,
Houston, Texas
July 25, 1969, 5:42 A.M. CDT
Sent from Escondido, California

CONGRATULATIONS, ON A JOB WELL DONE, TO YOU AND
YOUR CREW. EACH OF YOU HAS DONE A SUPERB JOB. AS
ONE AMERICAN TO ANOTHER MY HEARTFELT THANKS FOR
YOUR MOMENTOUS CONTRIBUTION

MAJOR JOHN W CARPENTER COLONEL USAF RETIRED
ESCONDIDO CA

John Carpenter was one of Armstrong's principal squadron mates during the Korean War. He was not a naval aviator like Neil but rather a pilot with the U.S. Air Force. As an air force major, Carpenter joined Neil's fighter squadron, VF-51, on an air force–navy exchange program. In the first months of Neil's combat service in Korea in late 1950, Neil usually flew as Carpenter's wingman, with Major Carpenter heading VF-51's sixth division under section leader John Moore. When Neil's F8F Panther jet was badly damaged in a mission over a North Korean target in September 1951 and Neil was forced to eject, it was Carpenter, his division head, who stayed with Neil until he ejected in the vicinity of an airfield near Pohang (designated K-3), located far down the coast of South Korea and operated by the U.S. Marines. Armstrong and Carpenter stayed in touch for many years, both attending regular reunions of Fighter Squadron 51.

"IN RECOGNITION OF THE WORLD'S
MOST FAMOUS ROCKHOUNDS"

Western Union Telegram
To: "Neil Armstrong"
Received Central Mail Room, NASA Manned Spacecraft Center,
Houston, Texas
July 25, 1969, 6:13 A.M. CDT
Sent from New Orleans, Louisiana

IN RECOGNITION OF THE WORLD'S MOST FAMOUS
ROCKHOUNDS IT IS THE GREAT PLEASURE OF THE GEM
AND MINERAL SOCIETY OF LOUISIANA INC TO BESTOW
HONORARY MEMBERSHIP TO YOU.

G L YOUNG PRESIDENT

"YOUR STUPENDOUS ACHIEVEMENT"

Western Union Telegram
To: "Apollo 11 Astronauts"
Received Central Mail Room, Manned Spacecraft Center, NASA,
Houston, TX
July 25, 1969, 6:41 A.M. CDT
Sent from Beverly Hills, California

YOUR STUPENDOUS ACHIEVEMENT HAS CAPTURED THE
IMAGINATION OF THE WORLD AND GIVES NEW SPIRIT
TO MEN AND WOMEN EVERYWHERE. BOTH PERSONALLY
AND AS THE DIRECTOR OF THE PEACE CORPS I EXTEND
CONGRATULATIONS FOR YOUR BREAVE FEAT AND THANKS
FOR YOUR SAFE RETURN TO EARTH.

YOUR FLIGHT TO THE MOON BOLDLY DEMONSTRATES THE
ABILITY OF MAN TO SET DIFFICULT GOALS ALMOST BEYOND
HIS REACH AND THEN BY TREMENDOUS EFFORT ACHIEVE

THEM. FOR WE EARTH-BOUND, AND PARTICULARLY, I
THINK, FOR PEACE CORPS VOLUNTEERS, THIS HAS SPECIAL
MEANING TODAY IN A WORLD WHERE PROBLEMS ARE
SO VAST AND SOLUTIONS SO DIFFICULT. OUR AMERICAN
VOLUNTEERS ALSO TRAVEL TO FAR OFF PLACES TO TACKLE
NEW CHALLENGING PROBLEMS. THE SUCCESS OF YOUR
VOYAGE WILL, IM SURE, GIVE THEN A GREAT BOOST BOTH
AS AMERICANS AND AS VOLUNTEERS FACING ARDUOUS
TASKS.

YOUR ACHIEVEMENT ALSO GIVES ME ADDED OPTIMISM
IN THE EFFORTS OF THE PEACE CORPS TO ASSIST THE
NATIONS WITH WHICH WE WORK TO SET AND ACHIEVE
HIGH GOALS. NOTHING IS BEYOND MANS GRASP, YOUR
FLIGHT SAYS, IF WE ONLY WANT IT ENOUGH TO STRIVE FOR
THE SEEMINGLY IMPOSSIBLE. AND SO IT GIVES US HOPE OF
A BETTER LIFE IN PEACE FOR ALL MANKIND.
WELCOME HOME FROM ALL THE PEACE CORPS.

JOE BLATCHFORD
DIRECTOR PEACE CORPS

*Joseph Blatchford (b. 1934) was the third director of the U.S. Peace Corps,
appointed by President Nixon in May 1969, and a huge fan of the space program. The most prominent hanging on his office wall was a copy of* Earth Rise,
the famous over-the-Moon photo taken by the Apollo 8 crew in December 1968.

"MAY THE 14 MILLION SCOUTS ALL OVER THE WORLD"

Western Union International Telegram
To: "Neil Armstrong"
Received Central Mail Room, NASA Manned Spacecraft Center,
Houston, Texas
July 25, 1969, 7:35 A.M. CDT
Sent from Manila, Philippines

EXUBERANT CONGRATULATIONS FOR THE FIRST STEP
BY AN EAGLE SCOUT STOP A GIANT LEAP NOT ONLY FOR
MANKIND BUT HOPEFULLY FOR WORLD SCOUTING.
MAY THE 14 MILL ION SCOUTS ALL OVER THE WORLD
DERIVE ENTHUSIASM AND INSPIRATION FROM THE
EXTRAORDINARY MERIT BADGE YOU HAVE JUST GAINED.
CORDIAL REGARDS.

ANTONIO C DELGADO
MEMBER WORLD SCOUT COMMITTEE

*Antonio Concepcion Delgado (1917–1992) was a highly prominent Philippine
industrialist and civic leader who, like Neil, had a strong connection to the Boy
Scouts. Joining the Boy Scouts of the Philippines at age 15, Delgado attended the
4th World Scout Jamboree, held in Hungary, in 1933. Three and a half decades
later, in 1968, he became President of his country's Boy Scout organization.
Then at the 1971 World Scout Conference in Tokyo, Japan—which Armstrong
attended and at which he gave a talk to the Eagle Scouts—Delgado became the
first Asian to be elected Chairman of the World Scout Committee.*

*Neil, too, had been active in the Boy Scouts since boyhood, earning the Eagle
rank—and appropriately so, being the astronaut that flew the lunar module
Eagle down to the first lunar landing. He did not forget them during Apollo 11.
Not only did he greet from space the Scouts attending their national jamboree
(held that year in Idaho) while flying toward the Moon on July 18, 1969, Neil
also carried a World Scout badge with him onto the lunar surface—a memento
that he gifted to the Scouts after his return. The Boy Scouts of America presented
Armstrong with its Distinguished Eagle Scout Award (as it would do with Apollo
astronauts Jim Lovell and Charlie Duke, who were also Eagle Scouts) as well as
the Silver Buffalo Award, the latter for noteworthy and extraordinary service to
youth on a national basis.*

*There are hundreds of letters in Armstrong's papers at Purdue from Boy Scouts.
For several years thereafter he took the time to write letters congratulating boys
who had achieved the ultimate rank of Eagle Scout. But once his mailing address
was posted on the internet in the early 1990s and he was deluged with requests,
he could no longer write personal letters to Scouts. Neil determined that "congratulatory letters should be from people who know the Scouts personally, who know
what they've achieved and honestly want to congratulate them. When Scouts get
letters from political potentates that have actually been written by staff members*

and signed by an autopen, perhaps it impresses the individual getting the award and receiving that message, but it's the wrong message. It's just something that the Scouts don't do right."[10] *Nonetheless, the appeal for letters kept coming. In the first five months of 2003 alone, Neil received 950 letters asking for congratulatory letters for new Eagle Scouts.* "*Over the years I have done a lot of work on behalf of the Scouts,*" *Neil would relate in 2003,* "*but I have not done any of that in recent years.*"[11] *Much to the chagrin of the Boy Scouts of America, he no longer had an official association with them. As for congratulatory letters to Eagle Scouts, in the last years of his life he would only write them to young Ohio residents whom he personally knew.*

"MANKIND'S MOST INTREPID JOURNEY"

Western Union Telegram
To: "Neil A. Armstrong, Commander of Apollo 11"
Received Central Mail Room, Manned Spacecraft Center, NASA,
Houston, TX
July 25, 1969, 12:40 P.M. CDT
Sent from West Lafayette, Indiana

CONGRATULATIONS ON THE SUCCESSFUL COMPLETION OF MANKIND'S MOST INTREPID JOURNEY.

NOT ONLY HAS THIS FANTASTIC VENTURE ESTABLISHED NEW GOALS FOR THE COMMON EFFORTS OF ALL MEN, BUT YOUR COOL COURAGE, YOUR DEVOTION TO THE HIGHEST IDEALS AND PRINCIPLES OF SCIENTIFIC INVESTIGATION AND EXPLORATION, YOUR WILLINGNESS TO SEEK ENDS THE REWARDS OF WHICH WERE UNCERTAIN AT BEST, HAVE REQUIRED EACH OF US TO RE-ASSESS OUR OWN PERSONAL GOALS AND PURPOSES.

ALL OF PURDUE UNIVERSITY JOINS IN HAILING YOU AND YOUR GREAT CREWMEN. PURDUE, THE "MOTHER OF ASTRONAUTS", PROUDLY WELCOMES ONE OF HER "SONS" HOME.

WE LOOK FORWARD TO A VISIT FROM YOU AND MRS. ARM-
STRONG AT THE EARLIEST POSSIBLE TIME.

FREDERICK L HOVDE
PRESIDENT PURDUE UNIVERSITY

*Armstrong graduated from Purdue University with a bachelor's degree in aero-
nautical engineering in January 1955. He had started at Purdue in September
1947 on a four-year scholarship from the U.S. Naval Aviation College Program
(known as the Holloway Plan, after Admiral James E. Holloway, who had
fostered the program). After four semesters of schooling, the navy called him
to duty; he began flight training at Pensacola in February 1949. Following
his service as a fighter pilot with VF-51 in the Korean War, Neil returned to
Purdue in September 1952. At Purdue he played the baritone horn in the "All
American" Marching Band and pledged to the Phi Delta Fraternity. He sang
in the Purdue Varsity Varieties all-student review and wrote and codirected his
own short musicals.*

*Purdue University's engineering programs ultimately produced a significant
number of astronauts (twenty-three) but certainly none more prominent than
Armstrong. Naturally, in the decades following Apollo 11, the university would
cherish any association with its celebrated graduate. Over the years Neil would
contribute to Purdue in many ways. In the early 1990s he served as cochair
of a $1.5 billion university fundraising campaign with fellow astronaut and
Purdue alum Eugene Cernan, the last to set foot on the Moon, in December
1972, as commander of Apollo 17. Neil made many public appearances at the
university, including participating in a few football halftime events. On one
such occasion he was honored to beat Purdue's "World's Largest Drum" and wave
the university flag. He also agreed to let Purdue build the Neil A. Armstrong
Hall of Engineering, a 200,000-square-foot facility costing $53.2 million that
opened in 2007 and housed the School of Aeronautics and Astronautics, School
of Materials Engineering, and the School of Engineering Education, the first in
the country. For the front of the building Purdue commissioned a life-size bronze
statue of Neil as a young college student along with a replica of his footprints on
the Moon leading to the building's entrance. In 2008 Neil donated his papers
to Purdue, some 223 cubic feet (in more than 350 boxes) of material from his
personal files, including the letters, telegrams, and cards from which the items
for this book were selected.*

Frederick L. Hovde was the seventh—and longest serving—president of Purdue University, from 1946 to 1971.

"WE ARE LOST IN ADMIRATION"

Western Union International Telegram
To: "Apollo 11 Astronauts"
Received Central Mail Room, Manned Spacecraft Center, NASA, Houston, TX
July 25, 1969, 2:29 P.M. CDT
Sent from Geneva, Switzerland

CERN WHICH INVESTIGATES T SMALLEST PARTICLES OF THE UNIVERSE SENDS ITS SINCEREST CONGRATULATIONS TO NASA WHICH INVESTIGATES THE LARGEST PARTICLES ON HE TRIUMPHANT CONCLUSION OF THE APOLLO 11 MISSION WE ARE LOST IN ADMIRATION FOR THIS HUMAN AND TECHNOLOGICAL CHIEVEMENT BEYOND ALL PRAISE

BERNARD GREGORY
DIRECTOR GENERAL AND STAFF
EUROPEAN ORGANIZATION FOR NUCLEAR RESEARCH
GENEVA

Bernard Gregory (1919–1977) was a prominent French physicist who served as director general of CERN from 1966 to 1970. At CERN, Gregory supervised the construction of the world's first hadron collider, a large particle accelerator built to test theories in high-energy and nuclear physics.

"WHOLE WORLD REJOICES"

Western Union Telegram
To: "Neil A. Armstrong, Commander, Edwin E. Aldrin, Jr., Lt. Col. Michael Collins"

Received Central Mail Room, Manned Spacecraft Center, NASA, Houston, TX
July 25, 1969, 2:44 P.M. PDT
Sent from San Francisco, California

WHOLE WORLD REJOICES IN YOUR SAFE RETURN AFTER EPIC APPOLLO 11 MISSION TO MOON. SAN FRANCISCO HAS FOLLOWED YOUR FLIGHT WITH PRAYERS AND PRIDE AND NOW WISHES TO "OPEN ITS GOLDEN GATE" IN TRIBUTE TO YOUR MAGNIFICENT ACHIEVEMENT. YOU ARE INVITED AT YOUR EARLIEST CONVENIENCE TO SAN FRANCISCO FOR A TREMENDOUS TICKER TAPE PARADD AND CIVIC RECEPTION. YOU FULFILLED THE AGELESS AIM OF MAN WALKING ON THE MOON. SAN FRANCISCO SALUTES YOU.

SINCERELY
JOSEPH L ALIOTO MAYOR OF SAN FRANCISCO

Joseph Alioto (1916–1998) was the thirty-sixth mayor of San Francisco, California, from 1968 to 1976. At the turbulent 1968 Democratic National Convention in Chicago, Alioto delivered the speech nominating Hubert Humphrey, and there were rumors that Humphrey might select Alioto as his running mate (instead Humphrey selected Edmund Muskie.) Alioto presided over a time of turmoil and change in San Francisco.

"ABOVE ALL FOR YOUR HUMAN COURAGE"

Western Union International Telegram
To: "Messrs. Armstrong, Aldrin, Collins"
Received Central Mail Room, Manned Spacecraft Center, NASA, Houston, TX
July 25, 1969, 3:38 P.M. CDT
Sent from Zwevegem, Belgium

POPULATION OF BELGIAN TOWN ZWEVEGEM WANTS TO CONGRATULATE YOU WHOLEHEARTIDLY FOR YOUR

TECHNICAL PERFORMANCE AND ABOVE ALL FOR YOUR
HUMAN COURAGE AND SERVICE TO MANKIND

THE MAYOR OF ZWEVEGEM BELGIUM

*From King Baudouin I (1930–1993) to the simplest peasant in the Flemish
countryside, Belgium was a country seemingly infatuated with Neil Armstrong
and the first Moon landing. The king sent Armstrong several letters and telegrams,
and the number of letters from Belgian citizens to Armstrong was, over the years,
disproportionate to Belgium's population relative to other countries.*

"DINNER AT THE RAINBOW GRILL"

Western Union Telegram
To: "Neil A. Armstrong, Michael Collins, Edwin E. Aldrin"
Received Central Mail Room, Manned Spacecraft Center, NASA,
Houston, TX
July 25, 1969, 3:40 P.M. CDT
Sent from New York, New York

PLEASE EXTEND OUR INVITE TO ASTRONAUTS ARMSTRONG
ALDRIN AND COLLINS FOR DINNER AT THE RAINBOW
GRILL WHERE DUKE ELLINGTON WILL BE PLAYING WHILE
THEY ARE IN NEW YORK CITY WE WOULD BE HAPPY TO
ACCOMMODATE THE ASTRONAUTS THEIR FAMILIES AND
ANY SPACE AGENCY OFFICIALS WHO ARE TRAVELING
WITH THEM IN ONE OF OUR PRIVATE DINING ROOMS
THEN TAKE THEM INTO THE GRILL WHEN THE DUKE
ELLINGTON SHOW BEGINS. AS YOU KNOW MR ELLINGTON
WAS COMMISSIONED TO COMPOSE A MUSICAL TRIBUTE TO
THE APOLLO LUNAR LANDING AND HE WOULD BE MOST
HONORED TO PLAY IT FOR THE MEN WHOSE COURAGE
AND SKILL MADE THE LANDING POSSIBLE WE WILL DO
EVERYTHING WE CAN TO ASSURE THE ASTRONAUTS AND
THEIR PARTY UTMO PRIVACY. PLEASE RESPOND TO ME AT
THE ADDRESS BELOW

MARILYN AUGBURN
SOLTERS AND SABINSON INC
62 WEST 45TH ST
NEW YORK NY 10036

Marilyn "Kay-Kay" Augburn (1942–2012) was a native of Muncie, Indiana, and a 1963 graduate of DePauw University in Greencastle, Indiana. Early in her career she did administrative work for magazines, including the New Yorker, *and worked for the flamboyant publicist Lee Solters in New York City. She married Philip R. Sharp (D-Indiana, who served in the House of Representatives for ten terms, from 1975 to 1995). She made her own name writing Washington, D.C.–based political thrillers, including her 1979 novel* Sunflower *about the kidnapping of a U.S. president's daughter.*

"FOR CONDITIONS IN WHICH SCIENCE CAN FLOURISH"

Western Union International Telegram
To: "Astronauts Armstrong, Aldrin, and Collins"
Received Central Mail Room, Manned Spacecraft Center, NASA, Houston, TX
July 26, 1969, 7:56 A.M. CDT
Sent from Helsinki, Finland

ON BEHALF OF WORLD COUNCIL OF PEACE AND ITS NATIONAL ORGANISATIONS IN OVER ONE HUNDRED COUNTRIES OF ALL CONTINENTS WE SEND YOU HEARTY CONGRATULATIONS ON YOUR HISTORIC VISIT TO THE MOON STOP BY YOUR COURAGE YOU HAVE OPENED A NEW CHAPTER IN MANS ENDEAVORS TO CONQUER SPACE STOP YOUR SUCCESS IS AN INSPIRATION TO ABL MANKIND TO WORK WITH GREATER VIGOUR THAN EVER BEFORE FOR WORLD PEACE AND FOR CONDITIONS IN WHICH SCIENCE CAN FLOURISH FOR THE PROGRESS AND HAPPINESS OF ALL MANKIND

ROMESH CHANDRA SECRETARY GENERAL

WORLD COUNCIL OF PEACE
LONNROTINKATU 18 HELSINKI FINLAND

*Romesh Chandra (1919–2016) was a leader of the Communist Party of India.
He took an active part in the Indian independence struggle as a student leader
after joining the CPI in 1939. After serving for over a decade as the secretary
general of the World Peace Council, Chandra became its president in 1977.*

"OUR JOY IS TINGED WITH A BIT OF SORROW"

Western Union Telegram
To: "Astronauts Armstrong, Col. Edwin E. Aldrin, and Lt. Col.
Michael Collins"
Received NASA Kennedy Space Center Communication Center, Cape
Kennedy, Florida
July 22, 1969, 1:34 P.M. EDT
Sent from Chicago, Illinois

ON BEHALF OF THE OFFICERS AND MEMBERS OF
THE NATIONAL ASSOCIATION OF NEGRO MUSICIANS
INCORPORATED WE REJOICE WITH THE REST OF AMERICAN
FOR THE SUCCESSFUL LANDING OF THE ASTRONAUTS OF
APOLLO 11 ON THE MOON JULY 20 1969 BUT OUR JOY IS
TINGED WITH A BIT OF SORROW THAT YOU DID NOT FIND
IT WITHINT YOUR POWER TO ALLOW PARTICIPATION BY
ONE OF AMERICAS OLDEST ETHNIC GROUPS NAMELY THE
AMERICAN NEGRO WHO IN A GREAT MEASURE HELPED TO,
IN NO SMALLY WAY, TO MAKE AMERICA STRONG ENOUGH
TO BRING THIS MOONLANDING TO ITS SUCCESSFUL
CONCLUSION

JOHN E WEBB EXECUTIVE SECRETARY
NATIONAL ASSOCIATION OF NEGRO MUSICIANS

The ambivalence that African Americans felt about the Moon landing and the entire U.S. space program is clearly evident in this letter from the executive secretary of the National Association of Negro Musicians. No doubt John Webb knew that there had been civil rights protests at Cape Kennedy and not just for the launch of Apollo 11, though that was the largest of the protest events. Reverend Ralph Abernathy, successor to the late Dr. Martin Luther King Jr. as head of the Southern Christian Leadership Conference and de facto leader of the American civil rights movement, marched with four mules and about 150 members of the Poor People's Campaign for Hunger as close as they were allowed to get to the sprawling spaceport on the eve of the Apollo 11 launch. "We are protesting America's inability to choose the proper priorities," said Hosea Williams, the SCLC's director of political education, who claimed money spent to get to the Moon could have wiped out hunger for 31 million poor people. Interestingly, however, just as the National Association of Negro Musicians saw the space program as a potential bridge between black and white America, Williams stood "in admiration of the astronauts," just as Reverend Abernathy himself "succumbed to the awe-inspiring launch," declaring, "I was one of the proudest Americans as I stood on this soil. I think it's really holy ground."[12]

"RELATE AMERICA'S TWO EXTREMES"

Western Union Telegram
To: "Neil Armstrong. Edwin Aldrin, and Michael Collins"
Received Central Mail Room, Manned Spacecraft Center, NASA, Houston, TX
July 26, 1969, 10:26 A.M. CDT
Sent from Hartford, Connecticut

THE EBONY BUSINESS MEN LEAGUE REQUESTS YOUR PERSONAL APPEARANCE AT A NATIONAL BLACK BUSINESS SHOW AT THE NATIONAL GUARD ARMORY HARTFORD CONN 5 PM SATURDAY SEPTEMBER 1969. THIS WOULD PROPERLY RELATE AMERICA'S TWO EXTREMES OF THE EMERGING BLACK BUSINESS ESTABLISHMENT TO THE TECHNOLOGICALY OUT-OF-SIGHT AMERICA

RSVP
THEODORE M PRYOR
CHAIRMAN EXECUTIVE DIRECTOR

Theodore M. "Ted" Pryor (1921–2008) had a long career at Aetna Life and Casualty in Hartford, Connecticut. In the wake of the strides made by blacks during the civil rights era, he began to voice the need for economic development in the African American community. In 1966, he joined an effort to establish Connecticut's first black bank, Unity Bank and Trust, and in 1967 organized Hartford's approximately thirty black-owned businesses under an umbrella organization known as the Ebony Businessman's League. As its executive director, Pryor staged America's first National Black Business Exposition in June 1970, which showcased more than 130 black-owned businesses from all over the nation. Using the same points of persuasion as he raised in his telegram to Neil Armstrong, Pryor convinced Apollo 13 astronaut John Swigert to be one of the featured speakers.

"TAKE PRIDE ONCE AGAIN IN THE BRAVE, THE STRONG, AND THE FREE"

Western Union Telegram
To: "Neil Armstrong"
Received Central Mail Room, Manned Spacecraft Center, NASA, Houston, TX
July 26, 1969, 1:25 P.M. CDT
Sent from North American Rockwell Corporation Space Division, Downey, California

----

CONGRATULATIONS. THE ENTIRE MEMBERSHIP OF THE AMERICAN FIGHTER PILOTS ASSOCIATON JOINS ME IN SENDING OUR HEARTIEST CONGRATULATIONS FOR A SUPERB JOB. ALL AMERICANS CAN NOW TAKE PRIDE ONCE AGAIN IN THE BRAVE, THE STRONG, AND THE FREE. YOU HAVE PROVIDED US ALL WITH AN EMOTIONAL UPLIFT WE HAVE LONG NEEDED, AND WE SHARE WITH YOU THE

GLORY OF THIS GREATEST OF ALL ACHIEVEMENTS. I AM
GLAD I AM AN AMERICAN.

W M "BUD" MAHURIN
PRESIDENT AMERICAN FIGHTER PILOTS ASSOCIATION

*Colonel Walker Melville "Bud" Mahurin (1918–2010) was a U.S. Air Force
officer and pilot. During World War II, flying a P-47 Thunderbolt, he became the
first American pilot to become a double ace in the European theater, later becoming the only air force pilot to shoot down enemy planes in both the European and
Pacific theaters as well as during the Korean War. In May 1952 his F-86 was shot
down by North Korean ground fire; captured by enemy forces, Mahurin spent
sixteen months in a POW camp enduring brainwashing techniques. After his
release, he was assigned as vice commander of the 27th Air Division. He resigned
his commission in 1956 to accept a senior position in the aircraft industry in
Southern California.*

"STEEL PIER OFFERS $100,000.00"

Western Union Telegram
To: "Neil A. Armstrong, Michael Collins, Edwin E. Aldrin"
Received Central Mail Room, Manned Spacecraft Center, NASA,
Houston, TX
July 27, 1969, 7:53 A.M. CDT
Sent from Atlantic City, New Jersey

CONGRATULATIONS ON THE MOST SIGNIFICANT
ACCOMPLISHMENT IN THE WORLDS HISTORY. STEEL PIER
OFFERS $100,000.00 FOR ONE WEEKS APPEARANCE, LAST
WEEK OF AUGUST OR FIRST WEEK OF SEPTEMBER. PROFITS
MAY BE CONTRIBUTED TO CRIPPLED CHILDREN, HOSPITAL
OR CHARITIES OF YOUR CHOICE. APPEARANCE TO INCLUDE
ELABORATE BOARDWALK PARADE WITH OUR STATE AND
THE ENTIRE EASTERN COAST PARTICIPATING.

PLEASE CONFIRM.

GEORGE A HAMID SR
STEEL PIER

As the owner of Atlantic City's iconic Steel Pier, entrepreneur George Hamid Sr. (1896–1971) made major contributions to America's outdoor amusement and entertainment industry, staging fairs, circuses, carnivals, and expositions at the famous Atlantic beach resort.

"IT WAS QUITE A JOB TO DO"

July 27, 1969

Neil A. Armstrong
1003 Woodland Drive
Seabrook, Texas 77586

Dear Neil:

The job was done beautifully, Neil . . . and it was quite a job to do.

What a great adventure not only for you and the others who were there but for all of us who watched and got involved over the eight days you were gone. I don't believe that our outlook can ever be quite the same again toward space or even toward ourselves and our own world. Your part in the journey was magnificent so add my congratulations to what must be a whale of a pile.

Our trip to the Cape to see you off was very successful in large part because Jan took extra good care of setting things up for us. I can see that you did very well in choosing a wife. Kotcho and I took up where we left off years ago . . . right in the middle of situation comedy. We met them at the Orlando Airport where they had arrived an hour before and as Jad had warned Hertz "lost" my car reservation. I innocently asked to see the manager and because Kotcho had previously inquired using my name the desk gal got totally confused. In her confusion and frustration with me and the others obviously waiting she started to issue me a car by mistake. Then she quickly realized her mistake but by this time didn't want to admit it to me or her large audience. So we got a car. We could not possibly have planned a workable approach but in our stumbling . . . we did it.

As you may know we saw your sister and brother the night before at the party and had a delightful time catching up. I can't believe that your little sister whom I last remember as a nuisance at our patrol meetings can be that attractive mother of seven kids . . . but then she probably had some trouble with my graying hair.

Many thanks for the invitation. It was a great experience. Sorry to miss you and Jan. Would love to see you after all the ruckus dies down. It could be that the toughest part of the trip is the next two months. Let us know if you make the Boston area a stop.

Good luck to you Neil. Keep it up. Best wishes to you and Jan.

Bud
John Blackford
RFD 1
Concord, New Hampshire

John "Bud" Blackford and Konstantine "Kotcho" Solacoff were Neil's best friends during his early teenage years, when the Armstrong family lived in Upper Sandusky, Ohio. Bud, Kotcho, and Neil belonged to the same Boy Scout pack and Wolf Patrol within Troop 25. The three men stayed lifelong friends. Both Bud and Kocho were interviewed at length for First Man, *and a number of their stories about the days together with Neil, from 1941 to 1944, when they were boys of eleven to fourteen years old, are recounted in the book. As an adult, Blackford worked in the sawmill business in Contoocook, New Hampshire; Solacoff served as a family practice doctor in Upper Sandusky.*

"TURN YOUR MAGNIFICENT ACHIEVEMENT
TO THE GOOD OF MAN ON EARTH"

July 29, 1969

Astronauts Armstrong and Collins
Manned Space Flight Center
National Aeronautics and Space Administration
Houston, Texas

Gentlemen:

I am writing you in the pleasant capacity of spokesman for the Explorers Club to add a voice to the many millions of expressions that have come to you as the first explorers of the moon's surface. I also write in the unusual capacity of spokesman for a small group of men and women who are engaged in assessing the role played by a high mountain range of Alaska-Canada in the local and regional environment. These scientific explorers must surely be among the last people on this continent to have knowledge of the matchless success of Apollo Eleven for, while working between 2500 and 18,000 feet in the St. Elias Mountains, within our own communications network, we had no access to the radio and television coverage of the world at large. Mail reaching us today brings our first complete awareness of the journey that has brought unbounded credit to your team, to the backup that made it possible, and to the United States of America.

Throughout Man's history his kind has been endowed with a curiosity to understand what lies beyond his earthly horizon. Now the flight of Apollo Eleven has demonstrated that the unknown, while infinite, is within the bounds of man to explore beyond his terrestrial frontiers. In expressing to you the admiration of the Explorers Club, may I also venture to hope that, as leaders in space, you turn your magnificent achievement to the good of man on earth, and to the end that exploration of this planet, still far from complete, may be fostered in a generation of explorers now emerging in our schools and universities. Your example can serve only to unite mankind in the human understanding we all seek.

Most cordially,

Walter A. Wood
President
Explorers Club
46 East 70th Street
New York, NY

As written from:
A.I.N.A.—A.G.S.
Icefield Ranges Research Project
Mile 1054—Alaska Highway
Via Whitehorse, Yukon Territory

The membership of the Explorers Club has always been kept private, but it is known that all three members of the Apollo 11 crew have been fellows of the Explorers Club since the 1970s.

"CHILDREN TODAY NEED REAL HEROES"

Western Union Telegram
To: "Neal A. Armstrong, Isolation Chamber, NASA Houston"
Received Central Mail Room, Manned Spacecraft Center, NASA, Houston, TX
August 5, 1969, 8:30 A.M. CDT
Sent from San Ramon, California

LETTER FOLLLOWING WAS SUBMITTED TO THE SAN RAMON UNIFIED SCHOOL DISTRICT BOARD TONIGHT AND UNAMIOUSLY APPROVED. IN TRYING TO SELECT A NAME FOR THE NEW ELEMENTARY SCHOOL WE FIRST TRIED TO COMMUNICATE WITH THE COMMUNITY TO FIND THEIR DESIRES WE SENT LETTERS TO EACH HOME WHO HAD A CHILD THAT HAD BEEN IDENTIFIED AS WOULD BE ATTENDING THE SCHOOL AND WE HAD ARTICELES IN

THE CONTRA COSTA TIME VILLAGE PIONEER AND VALLEY
PIONEER NEWSPAPERS ENABLING ANYONE TO BECOME A
PART OF THE COMMITTEE ND TO HAVE AN OPPORTUNITY
TO VOICE THEIR OPINION AS TO A NAME. WE HAD YOUNG
PEOPLE THAT VISITED ALMOST EVERY HOME IN THE AREA
WITH A FORM GIVING EVERYONE A CHANGE TO WRITE
DOWN AN IDEA SO THE COMMITTEE FEEELS THAT THEY
GAVE THE COMMUNITY EVERY OPPORTUNITY.

TO VOICE A SELECTION WE ON THE COMMITTEE TO FIND
OUR SELECTION FIRST TRIED TO DECIDE ON A CRITERIA
THAT WE WANTED IN A NAME WE FOUND THAT WOULD
LIKE TO NAME THE SCHOOL AFTER OUR COUNTRIES
PIONEERS AS WE BEGIN TO LOOK AROUND FOR A NAME
WE FOUND THAT WE WISHES TO HONOR A LIVING
PERSON THHE PERSON THAT WE SELECTED HAS BLAZED
A NEW FRONTIER. HE BEGAN TO REALIZE THIS DREAM
BY RECEIVING HIS PILOT LICENSE BEFORE HE COULD
DRIVE A CAR HE RECEIVED HIS DEGREE IN AERONAUTICAL
ENGINEERING AT PURDEW UNIVERSITY AND THEN JOINED
THE NATIONAL AERONAUTICAL ADMINISTRATION. HE
BECAME ONE OF THE MOST ACCOMPLISHED TEST PILOTS
IN THE WORLD THE 7 YEARS HE TRAINED AT EDWARDS
AFB AND 1962 HE APPLIED FOR ASTONAUT TRAINING
AND WAS ACCEPTED HE WAS THE COMMANDER FOR THE
GEMINI 8 MISSION AND THE APOLO 11. WE HAVE CHOSEN
TO HONOR MR ARMSTRONG BECAUSE HE EXEMPLIFIES
ALL OF MANS PIONEERING FFORTS HIS FEAT STANDS FOR
ALL THE PEOPLE CONTAINING IN ANY WAY WITH NASA WE
FEEL THAT IN THIS NAME THERE ARE MANY THINGS THAT
CHILDREN COULD DO WITH IT. IT LENDS ITSELF TO ALL
ASPECTS OF A EDUCATION MATH SOCIAL STUDIES SCIENCE
AND READING. WE FEEL THAT THEY COULD HAVE A LOT OF
DIFFERENT EDUCATIONAL EXPERIENCES BY IDENTIFYING
THEMSELVES WITH THE SPACE PROGRAM BUT MOST
IMPORTANT OF ALL THE COMMITTEE FELT THAT CHILDREN
TODAY NEED REAL HEROES LIKE THE ABRAHAM LINCOLN
OF YESTERDAY TO IDENTIFY WITH AND TRY TO PATTERN

THEMSELVES AFTER THEREFORE THE COMMITTEE, THE
COMMUNITY AND THE CHILDREN WISH TO SUBMIT THE
NAME OF NEAL A ARMSTRONG ELEMENTARY TO BE USED
FOR THIS SCHOOL RESPECTFULLY SUBMITTED

MRS NORALEE CREZEE CHAIRMAN
9903 BRUNSWICK WAY
SAN RAMON CA

*Throughout the United States today, there are some two dozen elementary, mid-
dle, and high schools named after Neil Armstrong. Many places around the world
have streets, buildings, schools, and other places named after him. The number
of schools in the United States named after American astronauts counts well over
200. There are over thirty schools named after Christa McAuliffe (1948–1986),
the Teacher in Space from Concord, New Hampshire, who was one of the seven
crew members killed in the space shuttle Challenger accident.*

"THAT BREED OF AVIATOR THAT LOVED FLYING"

August 8, 1969

Dear Neil,

I wish to take this opportunity to add my congratulations to those of
the other millions who have been astounded by your spectacular performance in landing on the moon. Mere words are inadequate to fully
express our emotions as we followed the mission. Although everyone
was excited and thrilled over the voyage and the performances of you,
Aldrin and Collins, there was a very special excitement and personal
interest on my part because I had braved the "Wild Blue" with you years
ago (July–Sept. 1949) in an old SNJ, clipping along about 150 knots
while serving as your primary flight instructor at NAS Whiting Field,
Milton, Florida. Little did I dream that the blond, young flyer with the
quick, infectious smile would someday execute the wildest flight man
had ever dreamed of. I should quickly add that I claim no credit for your
success in aviation, unless it was to encourage the love for flying which

you already possessed when you entered the navy. You could fly when you were assigned to me, so primary was little more than a checkout in the SNJ (and you wanted jets!). I guess I did introduce you to aerobatics, and I'm sure you remember how I loved "C" stage.

I had many students who were well coordinated and mechanically expert at flying, but you had all of these qualities plus the one that placed you in a very special category, that breed of aviator that loved flying and everything associated with it, with a dedication and purposefulness which has sparked the improvements and advancements that have elevated flying to the exalted position it holds today.

You may recall why I have a good reason to remember how good a student aviator you were. I had another student who was about a month ahead of you, H. L. Bondurant. He was the student who forgot to lock his seat belt and shoulder harness on his second hop and fell out of the airplane. That was just the beginning of his difficulties. Anyway, I looked forward to my flights with you because I could relax and enjoy myself after a tense period with him.

I won't bore you further as I reminisce, but I just wanted you to express my sincere congratulations and best wishes to you on this tremendous occasion. I envy you and those who are still active. I was forced to retire in 1961 because of heart trouble, so I'm grounded and it hurts! I am still teaching. I'm chairman of the Social Studies Department of Chapman High School and accused by my associates as spending much of my time recruiting for the U.S. Navy.

My congratulations also on your fine family. Janet demonstrated an exceptional degree of finesse in fielding questions from the press during the mission. In fact, she displayed so much knowledge and familiarity with the mission that I couldn't help but speculate on her qualifications as a lady astronaut.

Neil, as you and your fellow astronauts are greeted on the tour which will follow your release from quarantine you will begin to understand just how proud all of us are of the Apollo 11 mission. Then you will know in part just how pleased and proud I am to have been around when you started on a career which actually led to the moon! I hope this is only another beginning for you and my best wishes for your continued success.

Sincerely,

Lee R. P. Rivers
Lieutenant Commander, UNN, Ret.
Apalachicola, Florida

Personal reply from Neil

September 25, 1969

LCdr Lee R. P. Rivers, USN (Ret.)
Apalachicola, Florida 32320

Dear Pal:

It was a great pleasure to receive your letter after so many years of lost contact. I certainly appreciate your very kind words concerning our recent adventure. It certainly was a satisfying experience for us and we are very encouraged by the reception it has been given around the world.

I remember the days at Whiting Field very well and the excellent beginning that you provide me. I also remember the occasion with Mr. Bondurant. I learned a lesson from his experience that isn't easily forgotten. I frequently see several other instructors of that vintage that you may remember—Bob Rostien, who is now a chief test pilot with Ling-Temco-Vought, and John Moore, who is on leave from North American Aviation where he has been a test pilot for a number of years and is now doing lecture tours around the country.

I recall vividly the evening I spent with you and your wife at your home near Milton and look forward to renewing the acquaintance. I frequently drive through Apalachicola enroute to the Cape and will certainly plan on giving you a ring on my next trip that way.

Sincere best wishes,

Neil A. Armstrong

"ALL EXPENSES PAID TO THE RUINS OF MACHU PICCHU"

Western Union International Telegram
To: "Apollo 11 Astronauts"
Received Central Mail Room, Manned Spacecraft Center, NASA,
Houston, TX
August 17, 1969, 12:00 A.M. CDT
Sent from Lima, Peru

CASPIA TOURS LIMA PERU CONGRATULATING THE THREE
GREATEST MEN OF OUR TWENTY CENTUERY WISHES
TO EXTEND TO THESE ASTRONAUTS AND THEIR WIFES
AN INVITATION TO VISIT WITH ALL EXPENSES PAID THE
GREAT INCA RUINS OF MACHU PICCHU AT THEIR EARLIEST
CONVINIENCE.

WITH GREATEST ADMIRATION
CASPIA TOURS

Armstrong received several invitations to accept all-expense-paid vacations and, in the months following Apollo 11, accepted a couple of them, including a trip to Acapulco, Mexico. As for visiting countries in South America, following Gemini VIII in the fall of 1966 he made (along with Gemini XI astronaut Dick Gordon) a twenty-four-day goodwill tour of Latin America, which included events in Panama, Colombia, Venezuela, Ecuador, Brazil, Bolivia, Argentina, and Peru. Although not on this trip, nor the one Caspia Tours offered, Neil eventually made it to Machu Picchu.

"BUCK ROGERS FINALLY COMING TO LIFE"

Western Union Telegram
To: "Apollo 11 Crew"
Received Central Mail Room, Manned Spacecraft Center, NASA,
Houston, TX
August 17, 1969, 10:00 A.M. CDT
Sent from Balla Cynwd, Pennsylvania

CONGRATULATIONS STOP WE ARE PROUD THAT YOU
SUCESSFULLY ARRIVED, LANDED, AND WALKED ON THE
MOON AND RETURNED SAFELY AGAIN TO EARTH STOP 35
YEARS AGO IN OUR HOME, ON JULY 21ST 1934 WE HEART
FIRST HAND FROM OUR FATHER OF THE ADVENTURES OF
BUCK ROGERS IN HIS FIRST FLIGHT TO THE MOON STOP
FATHER, PHIL NOWLAN ORIGNATED THE COMIC STRIP
BUCK ROGERS ON THAT DATE STOP WALKING ON THE
MOON WAS A COMMON PLACE TO US AT MAPLE AVE IN
BALLA CYNWD PA BUT IT WAS A THRILL AND EXCITING TO
ALL OF US TO WATCH YOUR FLIGHT INTO SPACE STOP BUCK
ROGERS FINALLY COMING TO LIFE STOP

OUR BEST WISHES
THE CHILDREN OF PHILLIP NOWLAN AND BUCK ROGERS

*American science fiction writer Philip Francis Nowlan (1888–1940) was
the creator of Buck Rogers. The character first appeared in Nowlan's 1928
novella* Armageddon 2419 A.D. *as the character Anthony Rogers. The comic
strip—illustrated by Dick Calkins—ran for over forty years, spinning off into
a radio series, a 1939 movie serial, and two television series. Nowlan and his
wife, Theresa Junker, had ten children: Philip, Mary, Helen, Louise, Theresa,
Mike, Larry, Pat, John, and Joe. The family lived in the Philadelphia suburb
of Bala Cynwyd.*

*One of the most memorable quotes associated with the early U.S. space pro-
gram was "No bucks, no Buck Rogers." It is one of the most memorable lines from
the 1984 film* The Right Stuff, *expressed by the actor playing Gus Grissom, one
of the original Mercury Seven astronauts; the line had appeared originally in
Tom Wolfe's 1979 bestseller, also titled* The Right Stuff. *Whether Grissom ever
uttered the comment is unclear, but it quickly became a popular space industry
adage. Its meaning has been interpreted in various ways, as "if you don't pay them
what they believe they're worth, they're not going to fly" and more broadly as "it
takes the excitement and adventure of human space flight to generate popular
interest (and therefore financial support) for space exploration."*

*Like virtually all of the other early astronauts, Neil Armstrong did not care
for* The Right Stuff *(especially the film) or the concepts behind it. Nor as a boy
was he an avid follower of Buck Rogers.*

"EXPECT YOU WILL BE LONGING FOR A NORMAL LIFE AGAIN"

PERSONAL
Mr. and Mrs. Neil Armstrong, NASA
Manned Space Craft Center
Houston, Texas

October 13, 1969

Dear Neil and Jan:

I can't tell you what a pleasure it was for Livia and I to see you both in Rio. It wasn't quite like old times—the moon was bound to make some difference—but it was a great relief to me to discover that neither of you had changed in the slightest. I expected no less of you both, but the cynical side of my nature had prepared me for the worst. Fame is a subtle poison that few can resist. I have no doubt that I would be the first to succumb, and am only glad I am unlikely ever to be tested. I know it must be great fun for a while, but by the end of this trip I expect you will be longing for a normal life again. I only hope you can find it.

The night at the Chalet Suisse followed by the Canacão was the most fun of all and quite like old times. The next day Livia went out and bought the recording of Maysa's show there for you, and Paulo, our young security man who is a familiar of the place, took it around to her together with your note. She was touched by your thoughtfulness in writing and autographed the record for you. I will send it along under separate cover (an expression that always sounds slightly racy to me).

I was quite serious about recording for posterity the scandals of the 1966 trip. It will be printed on rice paper, bound in buckboards, and illustrated with quaint woodcuts. It will be published in a limited edition (by the "Olympia Press") and distributed to a select group of subscribers consisting of you, the Gordons, and the Lows.

Our plans are still a bit vague, as always in this business, but we hope to be vacationing in Mexico between the 1st and 15th of December. Thereafter we will be enroute to California and could stop off in Houston for a night or so if you were going to be there. We would *not* stay with you—one family of four in a house is enough without adding another, and besides my five-year-old bombshell is quite capable of leading your sons from the straight and narrow, though she may be a bit subdued by that time on discovering not everybody speaks Portuguese. However it would be splendid seeing you so soon again, and renewing acquaintance with George and Mary R. and Barbara and Dick Gordon, even though the letter might be potted in a more literal sense than usual—that is confined like a specimen to a glass pot.

We should be leaving Port Alegre about November 25th, so if you could let me know before then if it looks possible I will go ahead and make tentative plans. By all means don't put yourselves out over this, but if it is possible it would be great fun.

Once again, it was splendid seeing you in Rio, and I hope it will be possible to repeat the experience in December. If not we will probably be in the U.S. for four or five years this time around, and should be able to get together with you in Washington or elsewhere in the not too distant future.

Our best to you both.

Ashley C. Hewitt, Jr.
American Consul
Consulate of the United States of America
Port Alegre, Brazil

Although this letter is congratulatory, it also previews a number of letters that appear in chapter 4, "For All Mankind," associated with the around-the-world goodwill trip (twenty-three countries in forty-five days) made by the Apollo 11 astronauts from late September to early November 1969. One of the tour stops was Rio de Janeiro, as the letter from Ashley Hewitt makes clear. (After his service in Brazil, Hewitt became deputy chief of mission-counselor at the U.S. embassy in Kingston, Jamaica [July 1973 to August 1975], followed by a posting with the Department of State as chief of the Industrial and Strategic Materials Division, Office of International Commodities, Bureau of Economic and Business Affairs.)

Hewitt's mention of George Low, Dick Gordon, and their wives refers to the twenty-four-day goodwill tour of Latin America that Neil, Janet, and the others made in October 1966, a few months after Neil's Gemini VIII mission (with Dave Scott) in March 1966. Gordon was just off his Gemini XI flight. George Low at that time was serving as head of Apollo Applications in Houston, having previously served as deputy director for manned spaceflight at NASA Headquarters.

Maysa is Maysa Figueira Monjardim, better known as Maysa Matarazzo, a popular Brazilian singer and actress, whose career featured Bossa nova music and the sultry interpretation of torch songs (fossa).

"YOU WERE A MOST PLEASANT SURPRISE"

November 28, 1969

Mr. Neil Armstrong
1003 Woodland Drive
El Lago
Seabrook, Texas

Dear Neil:

Now that you have ended your world tour and I am back from my own visits to Mar del Plata, Montevideo, Santiago, La Paz, and Caracas, I wish to let you know without delay that from all the activities of the last few weeks—among them our broadcast of the Apollo 12—my trip with you from Houston to Rio stands out as the most exciting event.

I must in all sincerity tell you that you were a most pleasant surprise to me. I could have easily forgiven you any sign of vanity or haughtiness. Yet you were a lesson in modesty and relaxed naturalness. I was deeply impressed.

Upon my return to Washington I was informed that the broadcasts I made from Air Force One were picked up by 2,251 stations in the Hemisphere. Besides during my stays in Mar del Plata, Montevideo, Santiago, and Caracas I rode on the Apollo 11 coat-tails handsomely. All the more important papers and networks requested interviews to know more about your flight and the personality of the astronauts. Needless to say I depicted you and Mrs. Armstrong with heartfelt and sincere admiration.

My children—Alvaro, Fernando, and Maria—were delighted to hear your voice on the tape you recorded for them. And I must thank you again for this gesture of sensitivity and generosity towards them because they will forever keep this recording as one of the most important souvenirs of their father's life.

Sincerely yours,

Enrique Gonzalez-Pegueira
United States Information Agency
Washington, D.C. 20547

"WEARING YOUR ACCOMPLISHMENTS WITH PRIDE AND POISE"

June 8, 1970

Dear Neil,

You sure have gone a long way since "Snow White and the Seven Dwarfs"!

I have proudly watched your tremendous accomplishments, Neil, bragging to my three sons all the way that I had known you in college. We moved to West Lafayette last summer where I am in school getting my teaching certification. When I took the boys out of school to see you at the Basketball Arena they couldn't understand why I didn't go up and talk to you. So you see I am writing this letter to save face. I presume you have some publicity available for hero worshippers. Here are three: Bob (8), Jeff (10), and Bart (13).

You are wearing your accomplishments with pride and poise, Neil. I wish you and your family the best in this life on Earth. I am proud to have known you—if but slightly.

Sincerely,

Jan Wakeman Colter
West Lafayette, Indiana 47906

While in college at Purdue, Neil became his fraternity's (Phi Delta Theta) musical director and played active roles in no less than two of the university's "Varsity Varieties" all-student reviews. In fact, Neil wrote and codirected short musicals for the shows held in 1953 and 1954. The first of his shows was, indeed, Snow White and the Seven Dwarves, *the show mentioned by Jan Wakeman (Colter), who performed in the show. For his second musical theater production, which he titled* The Land of Egelloc *("college" spelled backward), Neil took the music of Gilbert and Sullivan and wrote new lyrics for it.*

3

THE SOVIETS

Given the commanding strengths of Soviet ideology and the comprehensive character of its organs of state propaganda during the 1960s, one would assume that the people of the USSR and Eastern Bloc countries learned very little about the space achievements of the United States, their bitter Cold War rival. When the ambition of Kremlin leadership—and for that matter of the chief designer of the Soviet space program, Sergei Korolev (1906–1966)—was to beat the Americans to space "firsts" at all costs, sharing anything with Soviet citizens about U.S. efforts, other than America's corrupt capitalistic objectives and its failures to keep up with the USSR, made no sense.

But the reality inside Soviet society was not monolithic. A considerable number of its people so greatly admired the Apollo 11 mission and its astronauts that they found ways to send friendly, congratulatory messages—and kept sending them for years to come. Throughout the 1960s, Soviet propaganda had grown less and less convincing about its space achievements. By the end of 1966 NASA had completed ten highly successful Gemini missions, while the Soviets had not launched a single cosmonaut into space since Voskhod 2 in March 1965. The next four Voskhods were canceled, followed by the termination of Voskhod altogether. The first manned flight of Soyuz was not launched until April 1967 and not only would the flight of Soyuz 1 be plagued with technical issues but also it would end in tragedy when the descent module smashed into the ground due to a parachute failure, instantly killing cosmonaut Vladimir Komarov, the first in-flight fatality in the history of spaceflight.

In 1967 and 1968 the Soviets become less brazen in self-praise for their space missions, as no new firsts materialized for them. And though the American space program was hit in late January 1967 with the tragedy of the launchpad fire that killed the crew of Gus Grissom, Ed White, and Roger Chaffee inside their Apollo 1 spacecraft, the American program rebounded by the end of 1968 with the audacious mission of Apollo 8. In only the third launch of the Saturn V Moon rocket and the second manned flight of the Apollo program, the crew comprised of Frank Borman, Bill Anders, and Jim Lovell became the first humans to travel beyond Earth orbit, escape Earth's gravity, see Earth as a whole planet, enter the gravity well of another celestial body, orbit the Moon, see the far side of the Moon, witness an Earthrise, escape the Moon's gravity, reenter the Earth's gravitational well, and safely return to Earth. It was an astounding mission even the Soviets had to credit. Soviet spokespersons had no choice but to change their tune about racing the Americans to the Moon. On the day of Apollo 8's launch on December 21, 1968, veteran cosmonaut Gherman Titov, the second man to orbit the Earth back in August 1961, told reporters, "It is not important to mankind who will reach the Moon first and when he will reach it."[13] The day after Apollo 8's splashdown and safe recovery, Leonid I. Sedov—a leading member of the USSR Academy of Sciences (who many in the Soviet Union saw as "the father of Sputnik")—told Italian journalists that the Soviets were not even competing in a race to orbit or land on the Moon. Instead, the Soviet plan, at least publicly, was to move forward with designs for various auto-mated spacecraft. (As readers will see in this chapter, academician Leonid I. Sedov would write to Neil Armstrong in April 1973 inviting him to give a talk in the Soviet Union.) With "the propaganda of deflection" in place ("The Soviet space program is not racing the Americans to the Moon. It has higher goals than simply competition."[14]), it became more possible for the Soviets to give at least half-hearted credit to what the Americans were accomplishing in Apollo. Official Soviet doctrine began to report on American space activities and share a slightly more positive impression of them.

But touting Apollo 11's history-making mission was another mat-ter—not something that the Soviet system could abide. What Apollo 11 achieved was so breathtaking, such a zenith of everything that had been happening in space exploration, that it was very hard for the Soviet regime to credit it in front of their people at all. On Soviet television the only report of the news that the Americans had landed on the Moon came during a short break during a volleyball match between two local Russian teams, with no accompanying audio or visual footage. The only people in the USSR who saw any of the Western coverage of Apollo 11 was a select group of Politburo leaders, officers of the KGB, and men in charge of the Soviet space bureaus.

But the Soviet people were hardly enveloped in a total darkness. This has been made clear in the memoirs of Sergei Khrushchev, the thirty-four-year-old son of former Soviet premier Nikita Khrushchev, who was ousted from power in October 1964. In his published memoirs, Sergei, a career scientist, explained that it was impossible to keep such sensational news as the Moon landing from the Soviet people: "You cannot have people land on the moon and just say nothing. It was published in all the newspapers. There were small articles when Apollo 11 was launched. Actually, there was a small article on the first page of Pravda [the official newspaper of the Communist Party of the Soviet Union] and then three columns on page five."[15] Not that the Soviet people were bowled over by news of the Americans landing on the Moon. As Sergei Khrushchev explained, the mood of Soviet society was "very similar to the feeling among Americans when Yuri Gagarin, our first cosmonaut, went into orbit. Some of them tried to ignore it, some of them were insulted." Soviet propaganda "did not play it up or give too much information. I remember I watched a documentary on it; it was not secret, but it was not shown to the public. The Russian people had many problems in day-to-day life; they were too concerned about other things to be concerned about the first man on the moon."[16]

The Soviet people would have been much more concerned if their government had not kept from them the fact that their own space program was, in truth, rather desperately trying to beat the Americans to the Moon. Though official spokesmen strongly denied it at the time and for many years thereafter, the Soviet space program had spent many million rubles on a mammoth Moon rocket, the N-1, only for it to blow up on the launch-pad at the Baikonur Cosmodrome on July 3, 1969. Behind the catastrophic failure were years of bureaucratic in-fighting between the country's rival rocket and spacecraft design bureaus, all of which were competing for the same limited resources, people, missions, and mandates from the Kremlin. Then in a last-ditch effort to steal thunder from America's Moon landing, the Russians—on July 13, three days before Apollo 11 blasted off from Cape Kennedy—launched a small unmanned spacecraft, Luna 15, not just to land on the Moon but to scoop up a sample of lunar soil and return it to Earth before Apollo 11 got back. Aware of the launch from U.S. spy satellites, NASA officials worried that the actions of Luna 15 might interfere with Apollo 11—over the years that had happened occasionally when the Russians operated at or near NASA's radio frequencies. Ultimately, nothing about Luna 15 bothered Apollo, as the Soviet mission failed, with Luna 15 crashing into the Moon on July 21, hours after Apollo 11's successful landing. The Soviet space program quietly suffered another disgrace. It was hardly a good time for Moscow to be congratulating the Americans for Apollo 11.

Thus, though most Soviet people knew nothing about their own Moon mission failures, many thousands of people in the USSR and in associated socialist and communist states, learned a great deal about Apollo 11, thereby making it no surprise that Buzz Aldrin, Mike Collins, and especially Commander Neil Armstrong would receive dozens and dozens of letters and telegrams from people living in the Soviet Union and Eastern Bloc.

"I SEND YOU EFFUSIVE AND FOND EMBRACES"

[Translated from Spanish at NASA Headquarters]
To: "The North American Astronauts"
Received Central Mail Room, Manned Spacecraft Center, NASA, Houston, TX
July 23, 1969, 7:50 A.M. CDT
Sent from Havana, Cuba

I CONGRATULATE THE NORTHAMERICAN ASTRONAUTS ON A SUCCESSFUL LUNAR LANDING RENDERED MORE EXCITING BY A VISIT OF THE SOVIET FLEET AND BY THE SUCCESS OF APOLLO XI. FROM MY SOCIALIST COUNTRY I SEND YOU EFFUSIVE AND FOND EMBRACES, WITH THE DESIRE OF PEACE FOR ALL.

GUSTAVO MENDOZA AKETE
COMPOSER OF THE SOCIETY OF AUTHORS, CUBA

"ADMIRATION AND BEST WISHES OF THE SOVIET PEOPLE"

To: "Apollo 11 Astronauts"
Received Central Mail Room, Manned Spacecraft Center, NASA,
Houston, TX
July 23, 1969, 11:45 A.M. CDT
Sent from Washington, D.C.

FOLLOWING MESSAGE RECEIVED FROM FORMER VICE-
PRESIDENT HUMPHREY:

QUOTE: WARMEST CONGRATULATIONS ON THE
SUCCESSFUL CONDUCT OF YOUR HISTORIC MISSION.
THE ENTIRE WORLD SHARES OUR GREAT PRIDE IN YOUR
ACHIEVEMENTS. THIS MORNING, AFTER YOUR FIRST
WALK ON THE MOON'S SURFACE, I MET IN THE KREMLIN
WITH ALEKSEY KOSYGIN, CHAIRMAN OF THE COUNCIL OF
MINISTERS OF THE USSR. THE CHAIRMAN ASKED ME TO
SEND YOU HIS PERSONAL CONGRATULATIONS AND THE
ADMIRATION AND BEST WISHES OF THE SOVIET PEOPLE.
END QUOTE

ADOLPH DUBS
COUNTRY DIRECTOR
SOVIET UNION AFFAIRS

*The Kremlin's drive for hegemony in the Cold War had led to a heavy-handed
emphasis on space firsts. It was the Soviet launch of Sputnik 1, the first artificial Earth satellite, on October 4, 1957, that, in terms of global awareness,
gave birth to the Space Age itself. Other firsts came in such rapid and seemingly
logical succession that the Soviets grew accustomed to them—naively coming
to expect nothing less. A month after Sputnik 1, on November 3, 1957, the*

Soviets launched a second Sputnik, carrying a dog, Laika, the first animal to orbit the Earth. Luna 1, on January 2, 1959, became the first spacecraft reach the vicinity of the Moon. The year 1959 brought two more significant Soviet firsts: on September 14, Luna 2 became the first human artifact to make it to the lunar surface, and on October 4 (the second anniversary of Sputnik 1), Luna 3 took the first photographs of the Moon's far side. On February 21, 1961, the Venera 1 probe deliberately crash-landed on Venus after making the first flyby of another planet.

All of these achievements were touted internationally, adding enormous pres- tige to the Soviet Union. But the feat that stunned the world came on April 12, 1961, when cosmonaut Yuri Gagarin, aboard Vostok 1, became the first human in space. Announcing Gagarin's orbital flight by radio to the Soviet people was Yuri Levitan, the same broadcaster who had reported all the major victories over the Nazis in the Great Patriotic War. Instantly, Gagarin—the son of a carpenter who worked on a collective farm and a milkmaid—became a colossal hero. Throughout the USSR and Eastern Bloc, cities and towns staged mass celebrations so large and enthusiastic they were compared to the victory parades at the end of World War II. In Moscow, a long parade cheering Gagarin moved through the streets to the Kremlin where, in a lavish ceremony, before massive crowds and an assembly of all the highest government and party officials, Premier Nikita Khrushchev—famous in the United States for declaring "We will bury you"—awarded the cosmonaut the revered title Hero of the Soviet Union. It was not just the Soviets but the world that honored the first man into space; in the following months, the handsome cosmonaut would make high-profile visits to Egypt, Germany, Italy, Finland, Canada, Brazil, Japan, and even Great Britain.

The Soviet firsts continued. Aboard Vostok 6 on June 16, 1963, Valentina Tereshkova became the first woman in space, orbiting the Earth forty-eight times while spending almost three days in space; in that single flight, she logged more flight time than the combined times of all American astronauts who had flown to date. Sixteen months later, on October 12, 1964, Voskhod 1 became the first spacecraft to carry three people to orbit. Then, on March 18, 1965, the biggest achievement since the Gagarin flight occurred when cosmonaut Alexei Leonov made the world's first spacewalk, exiting his Voskhod 2 spacecraft and staying outside for over twelve minutes, connected to his craft by a seventeen-and-a- half-foot-long tether. Closing out the era of Soviet-dominated firsts was Luna 9, which on February 3, 1966, made the first soft landing of an object on the Moon.

None of these firsts were 100 percent successful in their design or execution. Some had major problems. Of course, much the same could be said about the early

American achievements in what was then called "manned" spaceflight—almost every mission had limitations and experienced problems, minor or major. The difference was that the Soviet government cloaked its spaceflights in near-total secrecy, publicizing only the most positive features of its successful missions and totally shrouding its shortcomings, failures, and disasters. Unmasked, as it would later be by historians, the visage of the Soviet space program was starkly different than the propaganda image.

"WE ARE OVERWHELMED"

[Translated from Russian at NASA Headquarters]
To: "Apollo 11 Astronauts"
From: "The City of Gomel"
Received USS *Hornet*
July 27, 1969, unknown time
Sent from Gomel, USSR

FROM THE CITY OF GOMEL TO THE COSMONAUTS OF APOLLO-11.

DEAR LUNAMEN, WE ARE OVERWHELMED AND VERY PROUD OF YOUR ACHIEVEMENT.
FAMILY OF MURTAZIN. USSR

"WE CONGRATULATE YOU FROM THE BOTTOM OF OUR HEARTS"

To: "The Crew of Apollo 11"
From: "The Cosmonauts of the USSR"
Received Central Mail Room, Manned Spacecraft Center, NASA, Houston, TX
July 29, 1969, 9:17 A.M. CDT
Sent from Moscow, USSR
Stamped "PRIORITY"

DEAR COLLEAGUES, ASTRONAUTS OF THE UNTIED
STATES—N. ARMSTRONG, E. ALDRIN, AND M. COLLINS!

WE SOVIET COSMONAUTS FOLLOWED YOUR FLIGHT WITH
GREAT ATTENTION AND EMOTION. WE CONGRATULATE
YOU FROM THE BOTTOM OF OUR HEARTS ON THE
COMPLETION OF THE MAGNIFICENT TRIP TO THE MOON
AND SAFE RETURN TO EARTH.

(SIGNED) G. TITOV, A. NIKOLAYEV, P. POPOVICH,
V. BYKOVSKIY, V. NIKOLAYEVA-TERESHKOVA, P. BELYAYEV,
A. LEONOV, K. FEOKTISTOV, B. YEGOROV, G. BEREGOVOY,
V. SHATALOV, B. VOLYNOV, A. YELISEYEV, AND YE. KHRUNOV.
AB/LOWCOCK/MUELLER

"BREATHLESSLY AWAITING AN ANSWER FROM YOU"

July 1, 1970 [Date received in Houston]

Esteemed Mr. Armstrong,

The words that you spoke as you stepped upon the surface of the moon,
"This is a small step for man but a large one for mankind," stirred me
deeply and I will never forget them.

The entire undertaking of the crew of "Apollo 11" is worthy of
admiration. It has denoted a turning point in the history of mankind,
modern and future generations, and will radiate admiration and rever-
ence of you to them.

In honor of this great moment I have decided, along with my friends,
to establish a rocket-astronaut club which will be named after you.

From the day the club was founded it started to grow and is gather-
ing more members daily. Various materials about astronauts has been
collected to insure the success of the club. We have obtained a movie
camera and projector, both 8 and 35 mm, unfortunately the latter size
film is very difficult to obtain in this country. In connection with this,

we would be most grateful to receive a few films about your travels in space which we will be willing to pay for upon receipt.

Due to the fact that we have connected you with the work of our club we would like to submit a few questions which will be useful in our newsletter which has been named "Apollo" and which is to be issued on 30 June 1970.

1. How much time did it take for all preparations for the Apollo 11 lunar flight?
2. How many hours or days have you logged in space?
3. Did you at any time feel that you would not complete your mission, that is, you would not descend to the moon?
4. How many space flights have you participated in and in what space craft?

Please accept our best wishes from all the members of the club who are breathlessly awaiting an answer from you.

Rocket-Astronaut Club "Neil Armstrong"
Vlasinska, Serbia
Socialist Federal Republic of Yugoslavia
President Zoran Stamenkovic
Secretary Dragan Stamenkovic
Vice President Perica Mirkovic

"THAT YOUNG FELLOW IN ROMANIA WHO DISTURBS YOU"

[Translated from Romanian by NASA]

July 21, 1970

Dear Sir:

On occasion of completing one year from the day when the first man, a terrestrial, has put his foot on another planet, and it was you, sir.

Whole-hearted congratulations to you, and I wish you many happy years and more success on other planets.

Esteemed sir, Neil Armstrong, I am C. Jase, that young fellow in Romania who disturbs you with my lines, this is the 4th letter which I have written to you, and I ask you to please excuse me if possible and I know whether . . . [illegible] even to please. I tell you sincerely.

Every day I waited for the mail, when will it come, believing that I can get it from you. Many thanks for the final letter which I received on the 5 of May. I tell you sincerely, I glanced over the letter several times, and I was very happy. But, at this time, I made the decision to play the loto every week, so I can win for you a small machine [model?] by time, and when you have a little free time, you would be willing to come to visit also my dear Socialist Republic of Romania, and you could admire its beauties, when you can come, and I with my family, will receive you warmly on the ground of my fatherland.

I close, and expect a new row of work. Let us compete much higher.

Sincerely,

Christian Jase
Ploesti, Romania

[Translated from Romanian by an unknown translator, July 1972]

[Unspecified date, July 1972]

Dear Mr. Armstrong,

I am that young man from Romania who would like to correspond with you, since I have had a single occasion to send a letter through RFG Tourist, so that it might be possible to strengthen my very great aspiration.

I would like very much to receive a model, even though used, to represent you in this country; but for me it would be a great source of pride and admiration, along with photographs of the first three earthlings to set foot on another planet.

I wish, as do all young people in my country, to view this model for a long time as the symbol of the friendship which you have for us.

And I also wish that this gift be sent through your embassy so that it would not be confiscated or taxed since my parents and I have no money. In all the letters I wrote you, I wrote of my family, my wife, but we are a poor family of workers.

I am proud that among the visits which you made in Europe you could come to my country so that we could see you personally, and not merely in photographs.

Many thanks for your kindness in reading my letters and for answering me and sending me photographs which remains for me waiting for your answer.

Sincerely,

Christian Jase
Ploesti, Romania

"HAPPY AND DEEPLY MOVED"

July 27, 1970

Dear Sir,

With the feelings both of concern and joy I was following the space-flight of Apollo 11 and I was happy and deeply moved when I saw you on the television screen standing on the moon surface.

I lived up to see this triumph of the man's ingenuity and courage in the 90th year of my life.

I would like to keep alive the memory of this historical feat by a document and therefore I take the liberty to ask you for your autograph. I shall keep it with greates respect for my children.

Will you please accept the expression of my most sincere gratitude and compliments.

Sincerely yours,

Alois Dyk
Prague, Czechoslovakia

P.S. Please use the attached sheet of paper.

"WE LOOKED WITH OPEN AND SCARED EYES"

July 30, 1970

Dear Niel Armstrong,

It is now more than one year when you, as the Commander of the U.S. spaceship 'APOLLO 11', landed for the first time on the moon on 21 July, 1969. One can never forget that auspicious moment when, after watching over the television throughout the sleepless night, early at the dawn, we looked with open and scared eyes how you took the first step on the moon. At that time you became the most glorious and the dearest man on the earth. This event is one of the greatest achievements in human history, as you rightly remarked, 'It is a small step by man but a great jump of mankind'.

Soon, on 5 August, the world will be celebrating your birthday, and I am exceptionally happy that on the same day I shall also be celebrating my birthday, together with the most celebrated person on the earth.

Please accept my most sincere and warm greetings on this joyous occasion, and also of my husband Dr. Vladomir Radovic and of my children Orhideja and Nikola. Also please convey our greetings to your wife and children.

With my personal best wishes,

Sincerely yours,

Mila Mirzova-Radović
(Dr. and Mrs. Mila Mirzova-Radović)
Zagreb, Yugoslavia.

"I RESPECT BESTLY AMONG ALL PEOPLE"

September 18, 1970

Dear Mr. Armstrong,

Please, excuse my taking Your time. I must admit, I often thought of writting You, especially after Your moon-landing. But I didn't do it, because I thought, You were very busy, and You wouldn't answer me. I daresay, You are occupied now as well, but maybe less. Thus dare ask You a great favour; write me please a short letter or card, some words only about Your life, Your experiences, or about Your future plans, all the same for me, the main point is that I should have a genuine document from the man, that I respect bestly among all people that live at present, from the man, that has taken the first footsteps on the moon. You know, I'm 17 years old, I live in Hungary (maybe You haven't heard of this small country), and my favourite passtime is to study the space flight, and the English language. You wouldn't believe, how much I know You from the foreign newspapers, and magazines.

Your handwritting would be my greatest wealth, and treasure. So please, answer me, if You can.

I'm looking forward to Your reply.

Yours very truly,

Jack Sebein
Budapest, Hungary

Reply from S. B. Weber, Neil's assistant

October 8, 1970

Dear Jack:

Thank you for your very kind letter to Mr. Armstrong. He is always pleased to hear from young people interested in our space program. Unfortunately, because of his heavy schedule of activities with NASA, he is unable to respond personally to all of his mail at this time.

We are enclosing, however, an autographed picture of Mr. Armstrong and an Apollo 11 Mission Report for your enjoyment. Best wishes and, again, thank you for writing.

Sincerely,

S. B. Weber

"I FEEL THAT WE ARE FRIENDS"

October 10, 1970

Honorable Sir:

I am in the 7th grade in the town of Celj. I watched your walk on the moon. You were very brave. I feel that we are friends. My father is dead and I live alone with my mother. I help my mother with the work. Would you please send me photographs of the astronauts and photographs from the moon. Thank you. Please accept my best regards,

Please extend my warmest regards to your family.

Majda Boncina
Gamilsko, Slovenia
Yugoslavia

"VERY MUCH LIKE TO KNOW HOW TO SPEAK ENGLISH"

October 16, 1970

Neil Armstrong,

I congratulate you on your courage and that of Collins and Aldrin. I watched with great pleasure as you exited from the module, and how the first "Earthman" stepped upon the surface of the moon. "This is a small step for me, but a large step for mankind" is the words that you spoke on the surface of the moon. I would very much like to know how to speak English, however we are learning Russian here, my father is the professor of the Russian language. I have never mentioned this to him for I am sure that this would anger him.

I congratulate everyone who has done such a good job.

I have a sister, a mother and a father in my family, and would like to extend greetings to the members of your family and I would also like to express my desire to correspond with your son or daughter, that is if you have children in your family. I would very much like to correspond with you but I know that you have very little time for this.

Goodbye

Omer Bacevac
Tutin, Yugoslavia

"I ESTEEM YOU IMMENSELY"

October 24, 1970

Respectable Comrade!

I esteem you immensly and therefore I am sending you this letter with one wish. The science which you are dealing with in question. Very soon I shall finish the eight grade school (gymnasium) and I would like to continue in a higher school where I could study the astronomy. I have great interest in this science from my early childhood and I wish to realize my dreams.

Please do write to me and tell me after which Yugoslav High-school I could continue to study in your country astronomy and whether it would be possible. I think that this science will realize to me a new direction in my life.

Please answer me in Serbian language.

I thank you in advance!

Dusan Veselinovio
Banjaluka, Yugoslavia

"THERE IS AN OLD LEGEND IN MY COUNTRY"

December 8, 1970

Dear Neil:

Let me first introduce myself. I am fourteen years old girl living in Poland near Warsaw. I have very good record at school. It is my seventh year of learning and I am continuously calssifierd on the top place on the list. My Daddy is an scientist and in the furture[,] I would like also to make the investigotinos of the matter[.] I was watching with greatest interest and anxiety the start of your rocket. The transmission form the landing on the Moon has take place in a middle of the night and I was allowed to watch TV overnight. Your first little step on the Moon surface has been indeed the big leap at mankind and I am proud that together with millions of people all over the world I was able to be an eyewitness of it. The pity is that that there is not too many such events nowadays that all people on all continents can be at the same time happy about it.

I remember well also the third mission to the Moon when Hus terrible explosion occured. I had been praying to Hie God for safe return of your friends on the Earth.

There is an old legend in my country from the times of King Sigmundus about a powerful magician Jan Twardowski who flew on the Moon. However he hasn't had so advanced technique as you because

instead of rosket he used the cock's back. But he didn't returned form
the Moon yet. Certain, your Apollo is much better vehicle than cock[.]
I am collecting photographs of the astronauts. I have got signed photographs at all your collegues form Mercury programe together [illegible]
form Mr Glenn and Mr Carpentier, also those of Russian (and Valentina
Tierieszkowa of course)[.] I would be very gratefull for sending me
your signed photograph as well as those of your friends form Apollo
programme.

Let me express the best wishes of many successful mission in the
exploration of the Universe and also ot Merry Christmans and kappy
pacetrill New Year

With love

Maigorzata Grabska
W. Parkowa, Poland

"FOR YOUR NOBLE TEARS, FOR YOUR GOOD HEART,
FOR BEING A REAL HERO OF MANKIND!"

[Translated from Russian at NASA Headquarters]

April 24, 1971

Dear Neil Armstrong:

May I address you with the most sincere good wishes for your health
and a long life for you and your wonderful family.

The tenth anniversary of space flight of our beloved and unforgettable
Yuriy Gagarin was recently noted. Gagarin, who loved the entire world!

Many newspaper articles, feature stories and recollections were published on this date.

One of the articles deeply touched our small family, my wife Nina,
daughter Natasha and myself.

In that article the author told how you, Neil Armstrong, met with
the two widows of our cosmonauts, Gagarin's wife and Komarov's wife,

and how you, a man of unlimited valor, whose courage no one in the universe can question, broke out sobbing! You could not speak . . .

Dear Neil Armstrong!

I thank you for your noble tears, for your good heart, for being a real hero of mankind!

I am writing this letter to you on the day when our remarkable lads [Vladimir] Shatalov [b. 1927], [Aleksei] Yeliseyev [b. 1934] and [Nikolay] Rukavishnikov [1932–2002] have returned from their [Soyuz 10] flight into space. Still another step toward knowledge has been taken.

I thank them, I thank you, I thank all the good people who have glorified Reason by their feats!

If you are ever in our Leningrad, a most beautiful city, where the courage of its inhabitants became the standard for behavior during our joint years of testing, my family and I will be happy to see and embrace you.

With sincere affection and respect, with wishes for happiness

Boris Tochinskiy
Leningrad, USSR

P.S. I wanted to send you, as a small token, an interesting cigarette lighter from my large collection, but Nina said that you probably do not smoke. I am sending several postage stamps (you probably have some young philatelists in your family).

"A NEW STAGE ON THE MASTERY OF THE PLANETS"

[Translated from Russian at NASA Headquarters]

August 29, 1971

Dear Neil Armstrong:

I am writing this letter and hope to obtain your answer.

I am 17 years old. I am a student of the Leningrad Institute. With great interest, I have been following the successes of our country and of the USA in the winning of outer space and I am collecting books and documents devoted to space research. You personally are the first Earthman to stand on the surface of a non-Earth celestial body. This event opened up a new stage in the mastery of the planets by man.

I would very much like to add to my collectin on space your picture with your autograph. I would be very grateful if you will not deny my request but send your autograph.

Sincerely,

Gennadiy Meshcheryakov
Leningrad, USSR

"I HOPE AMERICA WILL GIVE ME A FREE HAND"

[Translated from Slovenian at NASA Headquarters]

October 17, 1971

Dear Neil Armstrong,

I am interested in being the commander of Apollo 16, with Armstrong as my lieutenant. I know [illegible] to 2/4 of the moon, and the sun and stars. I hope America will give me a free hand with my command. Everything is different than you think in America.

Please send your answer to my address in West Germany, where I am employed: Glasschüttenwerk, 8594 Fichtelberg, Oberfranken.

I hope we can get together as soon as possible, to learn my story.

Mr. Neil, which country did the rocket belong to that went across West Germany? I have seen it with my own eyes. Please let me know. It passed below the moon, the earth became dark, I looked up and saw this rocket.

If you cannot read this, please get a translator.

I have received your photograph. It is very nice and I like it. Thank you.

However, what is shown, is completely wrong. There is no such thing on the moon. I have seen the moon, sun and the stars, and know exactly what they are like. There is no dust on the moon. Everything is different from what you Americans show. Everything you show is not true.

Am interested in Apollo 16. I have squared it with my conscience, for 2 million dollars you can hire me as Apollo 16 commander. Lt. Armstrong, you can go with me as pilot, and then one more who will have to suit my taste. You can introduce me to him in America.

Franz Senar (born 8 June 1928)
Ljutomer, Slovenia
Yugoslavia

"WE LIKED VERY MUCH YOUR CREW'S ATTITUDE"

October 28, 1971

Dear Mr. Armstrong:

First of all I wish to send you my heartiest regards. I am student in the 7th Grade of the Elementary Eng. Janek Krasinski School in Walbrzych. Our grade was very much interested in the flight of "Apollo 11" in which you and your crew participated. We liked very much your and your crew's attitude. The flight of "Apollo 11" enlarged our stock of knowledge about the space and the moon. I collect photographs of astronauts and therefore I would appreciate if you would send me your photograph.

Regards,

Renata Sommer
Walbrzych, Poland

"HOW DO YOU USE YOUR TIME, MR. ARMSTRONG?"

December 23, 1971

Dear Mr. Neil Armstrong,

Considering that nowadays youth is audacious and starting from a generous aim, I Carol Roman, a Romanian journalist, address you, Mr Neil Armstrong, as an outstanding personality who by your prominent presence in the contemporary world represent a "moral point of support" for the younger generations of our days—my wish to answer the world inquiry organized by the Romanian newspaper for the young people "Scînteia Tineretulai" ["Scent of Youth"].

"How do you use your time. Mr Armstrong?"

I am sure that in your ample process of creation and activity, in order to assert yourself, you had to make great efforts, to work day and night, to turn everything you have to good account, I am sure. I am convinced that only by mobilizing efforts, according to a certain programme of life, solicited this epoch of continuous progress of science and technique with a highest, most efficient use of your time in a certain regime of life, you managed to place yourself in the social forum where you are today. Indicating the way you keep in check "the time machine," how do you organize your life? How do you alternate your preoccupation? How do you work? How do you train yourself? How do you enjoy yourself, etc.? You could suggest possibilities and spiritual premises for the young people wishing to assert themselves.

Mr. Armstrong, We should like you to account in your answer a fact, a thought, a happening, a situation, even a regret, which should have a character of confession. We should even be interested in a working-day of yours which shows your regime of life—all these should be inscribed within our preoccupation of pleading for a rational use of life since youth. You can refer to this subject generally or fragmentarily or to a certain hypostasis.

I know you are a very busy person. Still I hope you will give some of your precious time for the younger generations.

As we intend your answer to be included in a volume we should ask you to send us the text—irrespective of the length—till December the

25th this year. We should ask you to send also a photograph of yours and a short autobiographical note: the date of your birth, studies, and successes.

With best regards and highest consideration.

Carol Roman
Bucharest, Romania

"COURAGE AND IMPUDENCE TO BOTHER YOU"

March 16, 1972

Dear Mr. Neil Armstrong,

Although I know you get very many letters every day, I have still got some courage and impudence to bother. I beg your pardon and I hope you will read my letter through.

Who am I?

My name is Uno Arnold Viigand and I am 19 years old. I live in Estonia, in the USSR. I am a first-year-student or so-called "Freshman" at Tarter State University.

My hobby is collecting of autographs. I have been collecting them for five years already and there are very many autographs in my collection from well-known Men around the world. I have got a request for you. If it is possible, please send me your photo with authograph.

With best wishes

Sincerely yours

Uno Arnold Viigand
Tartu, Estonia
USSR

Reply from Geneva Barnes, Neil's secretary

April 19, 1972

Dear Mr. Viigand:

Mr. Armstrong has asked me to thank you for your letter and to wish you every success at the University. He was pleased to autograph the enclosed picture for you.

Sincerely,

(Mrs.) Geneva Barnes
Secretary to Mr. Armstrong

"'ARMSTRONGITE' IN YOUR HONOUR"

June 7, 1972

Dear Mr. Armstrong:

We have sent to the Commission on new minerals and new names of minerals: The International Mineralogical Association, Washington, our new mineral sustained and submitted to this commission by the Ac. [Academy of] Sc. [Science], USSR.

We would like you to favour us with your agreement of calling this new mineral (silicate of calcium and zirconium): Armstrongite in your honour: the first spaceman setting foot upon the Moon.

Please forward your answer to Dr. [Michael] Fleisher.

Best wishes and thanks in advance.

Sincerely yours,

N. V. Vladykin
Institute of Geochemistry
Irkutsk, USSR

Personal reply from Neil

June 26, 1972

Dear Mr. Vladykin:

I received your letter through Dr. Fleischer with a great deal of pleasure. I am very pleased to have my name nominated for inclusion in this new mineral name and grant permission with thanks.

Please convey my appreciation to your colleagues at the Institute of Geochemistry.

Sincerely,

Neil A. Armstrong
Professor of Aerospace Engineering
University of Cincinnati

Michael Fleischer (1908–1990) served from 1959 to 1974 as chairman of the Commission on New Minerals and Mineral Names of the International Mineralogical Society. After earning a PhD in physics from Yale University in 1936, he spent three years as a chemist at the geophysical laboratory of the Carnegie Institution. From 1939 until his retirement in 1986, he was a mineralogist and geochemist with the U.S. Geological Survey. An internationally recognized authority on mineral nomenclature, he coauthored the Mineralogical Record's Glossary of Mineral Species *and had a mineral, Fleischerite, named for him. In 1992 the Russian Academy of Natural Sciences named him an honorary member.*

"YOU WILL BE THE GUEST OF THE USSR ACADEMY OF SCIENCES"

April 18, 1973

Dear Mr. Armstrong,

On behalf of the Organizing Committee I have the honour to invite you to participate the XXIV Congress of the International Aeronautical Federation (IAF) to be held in Baku, October 7–13, 1973.

You will be the guest of the USSR Academy of Sciences for ten days.

Expenses during your stay in the USSR will be covered by the USSR Academy of Sciences. Unfortunately the Academy of Sciences is unable to cover your travel expenses.

I would appreciate very much your prompt answer.

In case you accept the invitation you may obtain visas beginning with September directly through the Soviet Embassy in your country without applying to INTOURIST.

Please, let us know the exact dates of your arrivals to Moscow and Baku.

Meanwhile I thank you in advance and offer you my most cordial greetings.

Sincerely yours,

Academician L.I. Sedov
Chairman of the Organizing Committee
International Aeronautical Congress
Moscow, USSR

Personal reply from Neil

October 1, 1973

Dear Chairman Sedov:

With a great deal of regret, I must report to you that it is no longer possible for me to attend the IAF Congress in Baku as I had hoped.

Insurmountable schedule conflicts here in the United States have arisen which seem impossible to overcome.

This technical meeting would have been professionally rewarding and personally enjoyable, particularly inasmuch as your Academy of Sciences had so kindly offered to act as host. I know from my previous visit how enjoyable that would have been.

Please convey my thanks to the Academy and my sincere regrets for not being in attendance. I send my very best wishes for the success of the Congress and my personal sincere thanks to you.

Sincerely,

Neil A. Armstrong
Professor of Aerospace Engineering
University of Cincinnati

Neil had accepted an earlier invitation to give a talk in the Soviet Union. Shortly after returning from the Apollo 11 mission, he had received an invitation from the Committee on Space Research (COSPAR), which had been established in 1958 (a few months after Sputnik) by the International Council for Science (CSU), an organization dating back into the early twentieth century (formerly named the International Council of Scientific Unions), which was devoted to international cooperation in the advancement of science. COSPAR's charter of 1958 mandated that the organization be led by coequal vice presidents, one from each of the two superpowers. This unusual arrangement opened a door to dialogue and informal contacts between American and Soviet space officials and helped lead, notably, to the Outer Space Treaty signed between the United States and the USSR on January 27, 1967 (ironically, the same day as the tragic Apollo 1 fire). The first vice president appointed by the Soviets (always at the Kremlin's blessing) was Anatoli A. Blagonravov (1894–1975), a member of the Academy of Sciences of the USSR, past president of the Soviet Academy of Artillery Sciences, and an expert on long-range ballistic missiles. In 1970 the Committee on Space Research would hold its thirteenth annual conference, for the first time in the Soviet Union, in the city of Leningrad. (The conference had been held behind the Iron Curtain twice before, in Warsaw in 1963 and in Prague in 1969; its inaugural meeting in 1958 had been in London, with a meeting in Washington, D.C., in 1962.) In a clear indication of the strength of interest in the achievements of Apollo 11, COSPAR invited Neil to present a paper at its assembly in Leningrad.

Armstrong attended. On May 24, 1970, he arrived at the Leningrad airport on a flight from Warsaw. A red carpet awaited him but no crowds, as the Soviet government had not released news of his arrival. But at the meeting hall the commander of Apollo 11 received a great welcome and was mobbed by Soviet and Eastern Bloc scientists seeking autographs. After five days in the city, Neil was given permission to visit Moscow. At the Kremlin, he met with Premier Kosygin for an hour. On behalf of President Nixon, Neil presented him with some chips of a Moon rock and a small Soviet flag that had been carried aboard Apollo 11. The next morning Kosygin sent Neil bottles of vodka and cognac. Andrey N. Tupolev, the great Russian aircraft designer, and his son Adrian took Neil to the airfield hangar where they kept their supersonic TU-144, the rival to the Anglo-French Concorde; Neil was, in fact, the first Westerner to see the Soviet SST. The Tupolevs gave Neil a model of the TU-144, which Andrey Tupolev signed. During the trip Neil met several Soviet cosmonauts. In a secluded forest outside of Moscow, he spent the day at the Cosmonaut Training Center, which was part of the space complex of Zvezdny Gorodok ("Star City"), Russia's version of Houston's Manned Spacecraft Center. His hostess was Valentina Tereshkova, the first woman to fly in space, whom he found charming. Neil toured their training facilities, simulators, and spacecraft mockups. Tereshkova took him to the office of the late Yuri Gagarin, killed in an aircraft accident in March 1968 and whose personal effects had been preserved as a shrine to the first human space traveler. Neil then gave a lecture attended by many of the cosmonauts. Afterward, Mrs. Gagarin and Mrs. Vladimir Komarov were brought up to meet Neil, as Neil and Buzz had left medallions on the lunar surface in their husbands' honor. There were some Soviet media at the event. That night the cosmonauts invited Neil to a private dinner. After much toasting, they presented him with a very nice shotgun inscribed with his name on the stock. After dinner, around midnight, one of his hosts, Georgy Beregovoy (1921–1995) invited him to his apartment for coffee. (Besides being a veteran cosmonaut, Beregovoy was a decorated World War II flying ace.) At one point Georgy turned on his television set. On it was a tape of the launch of Soyuz 9 that had occurred earlier in the day. One of the cosmonauts in that spacecraft was Andrian G. Nikolayev, the husband of Valentina Tereshkova, with whom Neil had spent the whole day without her ever mentioning anything about the launch. Bringing vodka out for toasts, Beregovoy smiled broadly and told Armstrong, "This launch was in your honor!"[17]

4

FOR ALL MANKIND

The first Moon landing was a shared global event that transcended politics and nations. It was so much more than an American achievement; it was an existential attainment in which every person on Earth shared. Pride in the Moon landing was a universal, which became directly palpable to the Apollo 11 astronauts during their thirty-eight-day, twenty-three-country Giant Step tour from September 29 through November 5, 1969. "Traveling around the world in the months after the flight," recalled Mike Collins, Apollo 11 command module pilot, "I was continually impressed by the fact that no matter where we were, the reaction was the same. Never did I hear, 'Well, you Americans finally did it.' It always was 'we,' we human beings drawn together for one fleeting moment watching two of us walk that alien surface." It was a "clear positive effect" of the first lunar landing—still undeniable today fifty years later, even though the universal spirit has unfortunately not been sustained many places on the planet with much vitality.[18]

During the Giant Step tour, between 100 million and 150 million people saw the astronauts, with an estimated 50,000 people worldwide actually shaking hands with them and some 10,000 receiving autographs. In the aftermath of the trip, Armstrong felt that Apollo 11 and the Giant Step tour truly did some good to unite the world. Speaking at a graduation ceremony at Wittenberg College in his native state of Ohio in the spring of 1970, Neil said, "More can be gained from friendship than from technical knowledge," quite an admission coming from the devoted aeronautical engineer, test pilot, and astronaut.[19]

This chapter features letters to Neil from people around the world, from immediately after Apollo 11 through the last Moon landing mission, Apollo 17, which concluded with a successful splashdown on December 19, 1972. (Additional such letters can be found in subsequent chapters.) In contrast to the goodwill messages etched on the Apollo 11 silicon disc (featured in the preface), the letters in this chapter primarily come from everyday people, from all ranks of their societies, many of them from young people. What they all have in common is the desire to share a bit of themselves with the first man on the Moon and, if lucky enough, to get something back from him—a card, a letter, a photograph, an autograph, a thought, a sentiment—something they likely would cherish for the rest of their lives.

There is no exact count of the number of letters Neil received over the years from people living outside the United States, neither the raw total nor those preserved in the Neil A. Armstrong papers collection in the Purdue University Archives and Special Collections. An informed estimate for the latter would be 75,000 items, counting all manner of communication including cards, letters, telegrams, and printed copies of email messages. Based on my review of this material, the countries from which the greatest number of communications came to Armstrong over the years were (in rank order) Germany, Canada, Australia, United Kingdom, India, France, Italy, Pakistan, Japan, Spain, Belgium, and Argentina.

Over the years Neil would often consult an encyclopedia to learn about the place from which a letter had arrived, or he would ask his secretary to make a notation about the area. Later in his life his two sons jokingly referred to him as "Mr. Google" for his broadly based command of diverse information and his use of the internet to learn about what he did not know.

"IT WAS A BIT LIKE BREAKING DOWN SOME MAGIC"

August 11, 1969

Dear Mr. and Mrs. Neil Armstrong:

As an artist I am very deeply influenced by the Apollo 11 flight. I do not admire it only as an achievement of mankind because it anyway would have happened one day, but mainly as a heroic deed of three men.

The whole journey might be technically even easy for the crew but there still was so much quite new and impossible to figure out in advance in it. It was a bit like breaking down some magic and that is the point which never will lost it's shine.

Altough each of you have got a high technical education and you therefore should completely rely on your equipments and mathematical calculations made for the flight. Nevertheless it is quite clear you still have the very same human feelings as anybody else.

The ability for controlling these feelings should be called the courage of modern times. This courage was needed on the way.

Also the part of your families before and during the flight could not be the easiest ones. Though your wives no doubt were made well aware in detail of the whole operation and it's theoretic security they never before had experienced such a sense of relief as then as the craft splashed to the ocean.

I can be so sure of this because I myself sometimes during the flight suddenly found myself hoping that the men would just survive.

Beeing dominated by above kind of thoughts and feelings I got an idea to create a series of jewellery showing moon landscapes. I gave the serie a name "Moonrocks".

To show my admiration I decided to present the first three "Moonrocks" pieces to the families of the moon conquerors. These pieces are executed in gold and will remain unique pieces.

Congratulating you and hoping you will accept my gift, I am,

Sincerely Yours

Matti Hyvärinen
Turku, Finland

"I REALIZED THAT THIS MUSIC SHOULD BELONG ONLY TO YOU"

September 22, 1969

To our Astronauts Neil Armstrong, Michael Collins, and Edwin Aldrin Jr.

This is a message sent to you from Buenos Aires, Argentine Republic. S. A. It is from someone who has lived together with all his family the great emotion of your heroic feat.

As physician and investigator I have always explored the unlimited Universe called Microcosmos, more ever I have spared time for two hobbies: Astronomy and Music.

With a telescope bought in New York many years ago I have been able to study minutely the visible face of the Moon. That is why I have been familiar with the sight of the "Sea of Tranquility" on which you have been the first to step.

On one of those days of your great adventure without expressively porposing it to myself I sat down to my little piano and began to give expression to the sentiments that filled my heart.

After listening to what I have composed I realized that this music should belong only to you.

The grave chords to the rhythm of march seemed to me to terpret the firm and strong decision you have shown to march ever forward towards the Unknown, the lighter and more cheerful parts interpret your joy on seeing, each time you left the dark sight of the Moon, the image of your distant planet and the inmense satisfaction of a mission fulfilled.

Excuse the imperfections you may find in it. It was imposible for me to avoid the trafic noise of Buenos Aires.

Nevertheless I want you to know when listening to this music that there is a family here, who like millions more in all over the world has felt together with yours the same anguish and anxiety as well as the great joy of having you in your homes again.

I am ever sincerely yours

Dr. Carlos Velasco Suárez

Universidad de Buenos Aires
Facultad de Ciencias Medicas
Instituto de Oncologia
Buenos Aires
Argentina

N.B. The tape recorder as used as I was composing the music and I have not wished to shorten it in spite of its duration so that you may hear the original composition.

I have added three compositions more dedicated to Mrs Armstrong, Mrs Collins and Mrs Aldrin, these are two songs and a "tango."

July 1, 1970

Dear Mister ARMSTRONG

The letter with this one should have been sent when you started on your voyage round the world.

I thought of handing it to you here in Buenos Aires, but it was not possible nor oportune.

I have let the days pass until you should be home again rested after your long journey and less in demand.

Possibly next May I may go to Houston to assist at the 10th International Congress of Cancer.

Perhaps I shall then have the great honour and pleasure of making your acquaintance personally.

I am yours sincerely

Dr. Carlos Velasco Suárez
Universidad de Buenos Aires
Facultad de Ciencias Medicas
Instituto de Oncologia
Buenos Aires

Buenos Aires was the third stop on the Apollo 11 Giant Step Presidential Goodwill Tour, on October 1–2, 1969.

"I ALSO BEG TO GOD THAT ALL YOUR BEST WISHES BECAME TRUE"

February 2, 1970

Dear Mr. Armstrong,

At this marvellous moments, we ask succes triumph and trust for the men that went to the moon, I ask to God many succesfull for the Astronauts and their marvellous country.

Mrs Rosa that always talk about the three white lillys and pray to God when you Mr Armstrong landed on the moon, and not only this I spoke I beg for the return of yours to the earth, for me it is a duty wait good news of the explorers, I ask to the sky as a mother that want the triumph and the return of her childrens. The people say that the Astronauts are much nervous, I say not, they have faith they have not afraid, I also beg to God that all your best wishes became true, for the luck of the Science people.

Maybe you forget me, but I always will be praying for the triumph of the Science.

With sincere regards.

Rosa Aban de Terroyza
Buenos Aires, Argentina

"MY FAMILY REGARDS YOU ARE OUR CLOSE COUSIN"

June 12, 1970

DEAR NEIL ALDEN ARMSTRONG, MY DEAREST BROTHER,

THANK YOU VERY MUCH FOR YOUR KINDNESS. I AM
SO DELIGHTED AND ALSO VERY GLAD TO RECEIVE THE
BEAUTIFUL PATCH. I AM VERY HAPPY TO READ YOUR VERY
NICE LETTER. I THINK OF YOU VERY MUCH. I THINK YOU
ARE VERY GOOD MAN, YOU ALSO KEEP ONE'S PROMISE
FOR YOU CAN REMEMBER ME WHO IS ONLY A SMALL THAI
GIRL AND IS NOT IMPORTANT. I KEEP THAT PATCH WITH
YOUR LETTER AND YOUR PHOTOES IN THE VERY BEST
WAY. I AM SO SORRY THAT I COULD NOT ANSWER YOUR
LETTER AND THANK YOU AS QUICKLY AS I WOULD LIKE
TO DO. BECAUSE I WAS ILL AND I KNOW THAT YOU ARE
VERY BUSY ABOUT APOLLO 13. I KNOW THAT YOU ARE ONE
OF THE MEMBER OF THE COMMITTEE, SO I AM AFRAID
THAT YOU WILL NOT HAVE TIME BECAUSE YOU MUST
WORK HARD. MY PARENTS, MY BROTHERS AND SISTERS,
MY TEACHERS AND ALSO MY FRIENDS ALL WERE VERY
PROUD AND PLEASED WITH ME. MY FAMILY REGARDS YOU
ARE OUR CLOSE COUSIN.
  MRS. BARBARA DREXLER PEACE CORPS VOLUNTEER, MY
AMERICAN TEACHER AT SCHOOL WAS VERY GLAD, PROUD
AND ALSO VERY VERY INTERESTING, FOR MY FRIENDS
AND I HAD A CHANE TO MEET AND TO TALK WITH YOU.
SHE ASKED US TO WRITE A LONG COMPOSITION CALLED
"THREE THAI STUDENTS MEET THE ASTRONAUT NEIL A.
ARMSTRONG AT ERAWAN HOTEL" IN THE MOST WELL-
KNOWN COLUMN OF "SPACETIME," THE NEWSPAPER
OF OUR SCHOOL. ALL THE STUDENTS WERE VERY
INTERESTED IN OUR COMPOSITION. THIS COMPOSITION
MADE US BECAME WELL-KNOWN. THEY CALLED ME
"LUXIE" SINCE I USED THIS NAME IN THE NEWSPAPER.

I AM VERY GLAD THAT YOU GIVE ME A BLESSING. NOW YOUR BLESSING HAS COME TRUE, I HAD PASSED THE FINAL EXAMINATION AND ALSO PASSED THE EXAM FOR STUDYING IN THE UNIVERSITY IN "POLITICAL SCIENCE." IN SPITE OF THE FACT THAT IT WAS VERY DIFFICULT AND BUSY. ARE YOU GLAD?

I WISE YOU WERE MY REAL ELDEST BROTHER I LOVE YOUR SONS LIKE MY BROTHERS. WOULD YOU MIND LETINE ME BE YOUR SON'S PENFRIEND AND DO YOU STILL WORK ON APOLLO AT NASA ORGANIZATION? I ASKED YOU BECAUSE MY FRIEND TOLD ME THAT YOU DO NOT WORK ON APOLLO. IF IT IS TRUE, I WOULD BE APPRECIATED IT IF YOU TELL ME ABOUT YOUR NEW WORK AND NEW ADDESS.

I THINK THE APOLLO 13 DID NOT FALL AS SOMEONE SAID BUT I AGREE WITH YOUR PRESIDENT THAT IT IS NOT FAILURE, IT MUST BE THE GREAT SUCCESS OF NASA ORGANIZATION CERTAINLY.

I WISE I WERE A PERSON WHO WORK ON NASA BUT IT IS IMPOSSIBLE. I THINK I WILL GO TO SEE YOU AND YOUR FAMILY AT YOUR HOUSE SOMEDAY, IF YOU DO NOT MIND AND REMEMBER ME.

WOULD YOU BE SO KIND AS TO SPEND A LITTLE TIME, WRTIE TO ME WITH "A COLOURED PICTURE OF YOUR FAMILY." PLEASE GIVE MY RESPEST TO YOUR PARENTS AND YOUR WIFE, AND ALSO SAY HELLO TO YOUR SONS, "RICKY AND MARK." I WILL LOOK FORWARD TO RECEIVING YOUR LETTER AND YOUR FAMILY'S PHOTO. IN SPIRE OF THE FACT THAT YOU ARE VERY BUSY. BEST WISHES FOR CONTINEUED SUCCESS IN YOUR BUSINESS ESPECIALLY THE PROJECT OF APOLLO. TO BE SO PLEASED TO HEAR FROM YOU SOON, AND I HOPE YOU WILL BE STILL MY WELL-POWER FOR MY STUDYING IN UNIVERSITY

YOURS AFFECTIONATELY AND SINCERELY,

Luxana Nil-Ubol
Bangkok, Thailand

With the letter Ms. Nil-Ubol sent Neil a photo of herself, on the back of which she wrote: "To Neil . . . (Alden)—Best Wishes—Always be your sister and your friend—Yours sincerely & affectionately. Luxana Nilubol"

The Apollo 11 Giant Step Presidential Goodwill Tour stopped in Bangkok from October 26 to 28, 1969. This was in the middle of the Vietnam War, when many American military personal visited Bangkok for their R & R.

"I NEED A SMALL ROCKET"

June 15, 1970

Dear sir,

How are you? I live in Nicaragua in the Department of Rivas. I study in a National Institute and I am in the third year of secondary school. How does it feel to set foot on the moon and be the first human being there? I imagine that you were nervous and happy at the same time.

In my school we are having a 'nuclear fair' on space and the moon and I need a small rocket. Perhaps you could get one for me; it should have cameras inside to take photographs. I need this for the fair and I hope you will not disappoint me. Let me know how to launch it and how long it takes. I admire you all, especially those at the space center who have discovered a new world before the Russians.

I would like to study in the United States but I do not have the money. I am going to write to President Nixon, perhaps he will send for me.

I hope that you can send me what I ask for. My address is below.

Frank Alcaser Veliz
Rivas, Nicaragua

P.S. I hope for a favorable reply. Your friend.

"I MUST ADMIT THAT I ADMIRE YOU ABOVE ALL"

June 23, 1970

Dear Neil,

I can't quite think of anyone who has contributed so much to history as you have done by your success, as being the first human being to set foot on the moon. I realize a great deal of credit should be given to all the 'wonderful' people who have contributed to the exploration of space, as well, but perhaps fifty years from now, history books will concentrate more on you than on any other 'pioneer of space.' I think the reason for this is not only due to the actually mission but also to the results of your mission. For example: when you were descending from the 'Eagle' and was about to take that "one small step for man and one giant leap for mankind," the world watched this achievement in silence. At that time was united as one, and this is a great achievement within itself.

Though I admire many great people whether they lived in the past or live in the present, I must admit that I admire you above all. For this reason I would like to ask you, if I may please have the honour of having an autographed picture of you, to keep as a memento of the greatest man in history. I would appreciate it very much if you would please send it to my address which I have written at the top of this letter. Thank you very much for your time and consideration.

Sincerely

Loretta Busato
Toronto, Canada

Regrettably, the Apollo 11 Giant Step Presidential Goodwill Tour did not make a stop at any Canadian city, despite the fact that Canadians made several significant contributions to the Apollo program. Some of those contributions were provided by the University of Toronto Institute for Aerospace Studies. Perhaps Canada's most significant contribution to the U.S. space program came following the Canadian government's cancelation of the Avro Arrow, an advanced supersonic fighter plane being built in Malton, Ontario, a suburb of Toronto.

When the group of engineers that designed and built the Arrow was broken up in February 1959, a number of them went to work for NASA, where they helped form the core of the team that took America to the Moon.

"I HOPE YOUR CHILDREN ARE AS PROUD OF YOU AS MUCH AS I AM"

July 4, 1970

Dear Neil,

I hope you do not find me rude to use your first name, as it is the first time I have wrote to you.

I need'nt tell you how proud and thrilled your moon landing made me feel. (I have no doubt that you felt the same way.)

I watched and thourghally enjoyed the moon landing on T.V., (although I was so very tired.)

When I leave school, I hope to become a police woman. I will leave school in two years time when I will be seventeen.

I have always wanted to work with police, ever since I was about nine years old.

I cannot give you a description of myself, as I find it very hard to do so.

Regretfully, I have no recent photographs of myself, so I hope that you will take interest in this letter and next time when I write again, (if you don't mind) I will send a photograph of myself.

Since Apollo 10, I started doing a project. It has become very interesting with the help of Houston and London from where I recieved a tremendous lot of information.

I hope your children are as proud of you as much as I am.

My feelings for our bravery can be summed up in one word, "FANTASTIC."

Yours sincerely,

Sandra Mower
Saint Helens,
Lancashire, England

"PLEASE TELL ME HOW TO TRAIN AS AN ASTRONAUT"

July 7, 1970

Dear Mr. Armstrong,

I am a Japanese boy who go to senior high school. I am interested
in science.

Let me congratulate you on your success. 'Man tread on the ground
of the moon.' I was deeply impressed with the fact. We were watching it
on TV. I received histric paper from The New York Times.

What did you do while by the time you arrive the moon? Did you
feel solitary? If I were you, I won't be able to bear the solitude.

How do you think that you are the first man step on the moon
as mankind?

Please tell me how to train as an astronaut.

I am sorry to hear Apollo 13 failed. In America, is 13 supposed to be
unlucky, isn't it? In Japan, it is supposed to be unlucky, too.

When you came Japan, what was your impression of Tokyo? The
other day, Japan launched an artificial satellite. How did you think about
Japanese space development project?

Would you reply to letter? If you do so, I am happy.

Sincerely yours,

Manabu Saito
Mishima-shi, Shizuokaka
Japan

P.S. If you can, would you please send me your picture and
Nasa's sticker?

"BY NOW I EXPECT YOU ARE HEARTILY SICK
OF EARTH TRAVEL AND SPEECHES"

July 7, 1970

Dear Col. (?) Armstrong,

Forgive me if I have demoted you in rank, but write-ups over here have mentioned only your name.

My family and I have for many weeks wished to write and congratulate you on your most wonderful achievements in space, but it would appear you have been overseas most of the time, and we felt it would be wiser to wait until you had a more permanent address.

By now I expect you are heartily sick of earth travel and speeches, and will welcome the return to a more normal life if, indeed, you are now in Washington, as were lead to believe.

It is difficult to put into words our feelings for you as you stepped onto the moon's surface for the first time. I can only say it was so awe-inspiring that it was almost unbelievable, and that we who sat glued to our T.V. screens had such concern for your safety, and that of your fellow-Astronauts, that for those tense hours you became a part of the family.

We applaud your decision to quite flying, because we feel that after such a magnificent achievement for mankind, nothing should prevent you from enjoying a long and happy life with your lovely wife and family.

May I ask one tiny favour? That you will autograph the enclosed card, which we shall frame, and which will then become one of the most treasured possessions in our home.

Most sincerely,

Mrs. Marcia Langford
Killarney Heights,
New South Wales, Australia

"WITH ALL MY HEART, I LOVE YOU AND RESPECT YOU"

July 10, 1970

Dear Armstrong,

With all my heart, I love you and respect you. When you were in Apollo 11, I felt as if I was with you all the way to the moon. Every day, as soon as I get up from bed, I look at your photo nailed on the wall in front of me and gather strength and courage for my day's task. Inside my shirt pocket, I have another photo of yours, which I gaze at when alone or away from home.

Oh how lovely you are Mr. Neil, the bravest and the most courageous of men! I hail this day 20-7-70, as a mark of your setting foot on the moon a year ago.

How eagerly I wish for your autograph!

Your ever remembering friend,

M. Gunasekaran,
Tamil Nadu, South India,
India.

"I FOLLOWED YOUR MOON-SHOT SO CLOSELY THAT I COULD NOT RESIST WRITING"

July 14, 1970

Dear Mr. Armstrong,

As you will probably have guessed from what I have written on the outside of this letter I am writing to wish you a happy anniversary on your first anniversary of returning from the moon. Please do not doubt your memory for not remembering my name as, I am sorry to say, we have never met but I followed your moon-shot so closely that I could not resist writing to you at this time. I find it hard to believe that it is only one short year since I saw you on television, step down from the "Sea King" onto the "U.S.S. Hornet". I will never forget watching you and the other two lunar-nauts, looking like ghosts in your special isolation suits, walking tired but triumphantly, over to the "Silver Caravan" with it's significant sign of "Hornet +3". I don't know how you felt bobbing about in that capsule before you were picked up but I felt sick just watching it on television.

Needless to say I watched the landing on the Moon itself. I was one of the last survivors, in fact. My younger brother, intent on "keeping up with the Joneses," said he would stay up too. As soon as 11 came and went he dropped off to sleep in his chair. Finally at about 3 p.m. our time, we caught our first gimpse of your leg waggling about so we decided to waken him up. His reaction was "Well thay have landed, I'm going to bed." What a life. From that you will deduce that he is not, like me, a fully-fledged Lunar-Nut. Something which has puzzled me ever since then is, did you ever get any of that beautiful "Hornet-3" which was beside the "Silver Caravan" when you landed?? This may seem a very futile question considering that it was baked in your honor and that of Mr. Collins and Mr. Aldrin, it seems a bit ridiculous to think of your not getting a piece but on the other hand, I don't see how you *could* get a slice considering that you left it outside when you went into the "Silver Caravan" and you were supposed to be in quarantine from then on.

Well I suppose I'd better stick to questions like "When does the escape tower of the Saturn V rocket fall away?" At least I have *some* hope of getting an answer to that one.

Yours Sincerely,

Michael Mc Mullin (14 years old)
Stranorlar
County Donegal, Ireland

After becoming an avid golfer in his middle age, Neil Armstrong made many visits to Ireland and Great Britain to play golf with his two sons and with friends.

"YOU'RE MY IDOL"

July 14, 1970

Mr. Neil Alden Armstrong,

I'm an Italian girl fifteen years old. I'm going to write you because it's just one year you has walked for the I [first] time in mankind's history, on the Moon. So I wish compliment you on your enterprise. I've written you twice but I'll never tire to tell you that you're my idol & for me you're the most important man on the world for your courage, intelligence, cleverness talent &, allow me to say you, for your beauty. I'll come in the United States in a few days (July 25) & I should like to come in Houston but it's impossible. I'll go in New York, Albany, Washington, Canton (near Wapakoneta, your birth-place) and Niagara Falls. So I'll be able to visit the most important nation on the world that has succeed in a great enterprise like that of "FIRST MAN ON THE MOON"
Again congratulations!

Lovely Loredana Sivori
Rome, Italy

The Apollo 11 Giant Step Presidential Goodwill Tour visited Rome for three full days, October 15–18, 1969. While there, the astronauts and their wives visited the Vatican and enjoyed a papal audience with Pope Paul VI in the Papal Library of St. Peters Cathedral. (During the Moon landing the pope had gone to Vatican Observatory at Castel Gandolfo to look at the Moon through a telescope, making public remarks and offering a statement in English.) While in Rome, they also attended an elegant party at the home of the famous Italian actress Gina Lollobrigida.

"A SYMBOL OF APPRECIATION AND ADMIRATION"

July 15, 1970

Dear Mr Armstrong,

This letter comes to you from afar, but it brings you news which indicates how close you, Mr Aldrin and Mr Collins are to us in thought. We certainly admire your courage and success and now my family and I have a humbe way in which we have made a symbol of appreciation and admiration to your great achievement!

Enclosed you will find two photo copies of a Tibetan Carpet which has the words "APOLLO 11" woven on it. The explanation for this has been written by one of my sons. (also enclosed)

After weaving this we made another similar carpet, but really did not know where we should sell them. Finally, one of my children said that we should write to you and ask if you are interested in buying it, because it does bear the symbol of your unique success! Therefore, I wonder if you are interested in buying it? If not, please could you help us by organizing an auction in either a club or wherever you think it best and try to sell the carpets. If you, Mr Aldrin or Mr Collins wish to buy one or more each, we shall be most happy to make more of such carpets.

I assure you that the carpets are of 1st class quality and will last you a very long time. The materials of these carpets are of very superior wool from Holland. For your information I herewith enclose a slip which bears the trade Mark of the wool.

As you can see in the photo the carpets have been woven very skilfully and if the photoes have been coloured, you will really see how beautiful and unique the carpet looks!

As explained above, the carpets bear two aims:—i.e. they pay tribute to your historic success and they are the hopes for our childrens' needs and education. Therefore, I hope you will help us in whatever way you can. Any suggestions you can give will be most welcome.

With our very best wishes and hope of hearing soon.

Yours sincerely,

Karma Wangchuck

Dharmshala Cant (H.P.)
India

Size of the Carpet. . . .Length 6 ft. 4 Inches. Breath 3 ft. 1 inch.
Cost price for one carpet . . . Dollars 475.00 (four hundred and
seventy-five only)

Reply from S. B. Weber, Neil's assistant

November 4, 1970

Dear Mr. Wangchuck:

Thank you for your very kind letter to Mr. Armstrong offering him the
opportunity to purchase the fine rug woven by you and your family.
Your gesture is certainly appreciated; however, Mr. Armstrong is not in a
position to assist you with the sale of this item.

 We hope you will enjoy success in this and your many other endeavors. Best wishes.

Sincerely,

S. B. Weber

"EVERYTHING THAT NASA DOES IS OF
GREAT INTEREST TO ME"

July 16, 1970

Dear Mr. Armstrong,

Perhaps in your country people are remembering what day today is, but
here, much to my sorry, not even one word was mentioned of the greatest event there ever was.

I want to say that I am remembering today and all it meant to you and your colleagues Buzz Aldrin and Mike Collins. Everything that NASA does is of great interest to me and I send you all my very best wishes for all future endeavours. I would appreciate it if you would convey to Buzz and Mike my congratulations on the first anniversary and not forgetting yourself.

My best wishes always to you all.

Yours Sincerely,

Margaret Smith
Ladysmith, Natal
Republic of South Africa

"SHE WAS CHRISTENED PETRA M. APOLLA"

July 21, 1970

Sir,

On the day of the greatest triumph of the United States and the human race my little sister was born at the same time when you set foot on the moon, so she was christened Petra M. Apolla.

My name Manfred Kolar, I'm 17 years old and have two brothers, aged 15 and 10, and am a pupil of the grammar school in Graz.

My greatest wish for my sister's second birthday would be Neil Armstrong's autograph. With this biggest wish of my life so far I turn to NASA and hope you will be able to fulfil it.

With many thanks in advance and the wish that each further step forward in the space-programme may be crowned with success,

yours faithfully

Manfred Kolar
Graz, Austria

"I MUST CONFESS TO BEING A LITTLE JEALOUS"

July 23, 1970

Dear Sir Armstrong,

I would like first of all to hope that my letter reaches your hands and that when you have received it you will not tear it up and throw it away, if instead you have not done this and are now reading it, I thank you.

I have waited a long time to write to you until you had a little more free time than when you were so busy after your historic flight.

I am a neopolitan young man and justice in my shop my profession of hairdresser, but since child hood my hobby has been Astronautics which at that time was science fiction and which, thanks to you and thanks to your co-workers today has become reality in such a short time, so since I was small I have always saved and collected cuttings from news papers and magazines, books, recordings of the dialogue, of takeoff and splashdown, and of Transmissions from satellites; 8 mm films, stamps, coins, etc. all concerning space exploration, and I will always continue to collect because I think that this is the most beautiful thing that has happened in the world of Today.

My thanks go to you and your colleagues, who with your courage have made it possible for us to witness the most sensational undertaking of all ages—I must confess to being a little jealous to see you during the moon walk because the first man to set foot on the moon, I have been sincere in Telling you this and I will also tell you that I felt that also I was up there while I followed by Television that magnificent and stupendous spectacle, and throughout it all mission I often prayed for you and your colleagues. But I would not want to annoy you by speaking always of me. I will say only that I have written because I want so much to know something from you reguarding your work, but written by you yourself, even though it is only a few lines, and then if possible I would like you to send me a photo of you in your space suit, autographed, if possible so that I can add it to the precious possesions in my collection. If you do this for me I will be very greateful and very happy.

I hope you will forgive me for having made you lost precious time and forgive the of this letter imperfections, due to my imperfect knowledge of english.

I thank you and bid you

Good by

Peretti Alessandro
Naples, Italy

P.S. (If you can, send me the address of Sir Michael Collins. Thank you)

"I FELT THAT MY WHOLE BODY WAS SHAKING"

July 27, 1970

TO: Neil A Armstrong

My best wishes for you that you are doing fine along, with your family, also sharing happy moments in your social functions. Today that we celebrate the first anniversary since man walked ont eh Moon, I decided to write to you to express my admiration, and congrulate the rest of the astronauts, that realized the greatest act of these days and a gigant step to the Universe. I am a young girl that likes to follow step by step all of the projects to the Moon.

The trip of Apollo 11 was to me something unbelievable and from the moment that you all lift the earth I lived a along with you those instants of worry and anguish because, that is how I felt, and all I could remember was your parents and how they filt, that their son, was realizing something difficult, but at the same time great, and unbelievable. In those days our country, was in a difficult moments, for we had a war with El Salvador, but this wasn't enough for us to be pending, of your walk on the Moon, when this happened. I felt that my whole body was shaking, I watched the Moon at that moment but I couldn't see anything, but I knew and was sure that, there were two personas on that satellite.

Now every night watch the Moon, especially when it is full and it seems so impossible that it has, been visited by two representatives of the human nation. And I ask myself what do you all feel when you look at it and can say "One day I was there."

Not so long ago in a newspaper of this country came out a picture of your parents with a Honduran, that studies in that city donating same thing for his study, at the same time your mother said that if you where Invited buy the gobernor, of this country, you will have the opportunity to visit here, this is something that I wish for we'll have the opportunity to meet your personally.

I will like for you to share this letter with your parents, especially your mother, whom I admire for all the difficult moments. I will like very much to be your friend although through correspondence especially with your Mother who is proud each day of her son.

I will appreciate much if you write back to me and send your parents, address for I will like to be their friend, even though from this for country, I remember them and admire you, hoping to you very soon.
Best of luck always

Sincerely

Lexcy C Romero
La Ceiba, Honduras C.A.

"YOU STILL REMAIN AN ALL-TIME GREAT WITH ME"

July 30, 1970

Dear Mr. Armstrong,

I hope that this letter reaches you as close to the 5th of August (in your time zone) as possible, as I would personally like to send you my greetings and best wishes on the occasion of your birthday. (The 5th of August in America will be the 6th of August over here in Australia).

I am 15 years of age and am in my second last year of high school. I have always been a great admirer and follower of the Space Program, and your magnificent achievement over a year ago has made you into my greatest idol from among your fifty former colleagues, the other astronauts in Houston, although they are all brave and heroic men.

It is now over a year since your magnificent exploits, and now although you are out of the public eye and have a nice quiet desk job, you still remain an all-time great with me. I realise the number of people who have tried to reach you or meet you over the last 12 months, but I thought that now that it is your birthday, I might be at least able to send you a small note of thanks for the great joy you brought to me in July 1969. In conclusion, once again sending you my best wishes. Happy Birthday Neil Armstrong.

YOURS SINCERELY,

Leonard Halprin
Bentleigh
Victoria, Australia

In the ensuing years Dr. Leonard Halprin would come to know Neil Armstrong personally and maintain a correspondence with him right up to the time of Neil's death in 2012.

"I THEN STAYED UP ALL NIGHT WATCHING YOU"

August 1, 1970

Dear Friend:

No doubt you think it odd to get a letter from a girl you don't know!

Good! Allow me to introduce myself. My name is Conchita Estremera and I am 14; I am one of your admirers. Here is something you ought to know: when I am sad I will tell you my problems, and when I am happy I will tell you of my joy; when I go on a trip I'll take your photo along so that you can "see" what the life of a Spanish girl is like.

May I use the familiar pronoun with you? I hope so. Personally speaking, I would not want you to use the polite form.

Here is why I am writing: according to the papers in Spain, 5 August is your birthday; best wishes to you, and I hope they don't come too late.

I was so impressed by your trip to the moon that I still think of it every day.

Are you Catholic? I am; so on 20 July last year I got up at 7:30 a.m. to go to Mass. We then went for a trip and came home in the afternoon. I then stayed up all night watching you and your companions go to the Moon. I stayed up a total of 28 hours without sleep.

All the best for now.

Conchita Estremera
Campo, Huesca
Spain

"THEY WERE BORN THE DAY YOU LEFT THE EARTH"

[Translated from French at NASA Headquarters]

August 11, 1970

Dear Mr. & Mrs. Armstrong:

I have sent you these cards because my two children are now one year old. As you know, they were born the day you left the Earth, so you are the parents of these two children. This is why I have written you three times already to give them some money. They need clothes and I am poor. Help me. I wish you and your family a long and happy life.

Paul Amissale
Abidjan, Ivory Coast

On the back of the photo: "Claude Alain Amissale and Claude Vincent Amissale. One year ago, they were born on the day of departure of the Apollo 11."

"I AM STILL PROUDLY KEEPING THE UNCASHED CHECK"

August 12, 1970

Dear Mr. Armstrong,

I have the honour of meeting you in person during your visit to Turkey on October 21, 1969 in Ankara, when you stopped by the European Exchange System store where I owned a small Turkish souvenir counter from which you had bought some souvenir items as reminder of Turkey. I am still proudly keeping the uncashed check covering your purchase as the most valuable asset I own, which I will pass it to my sons and grandsons.

I am enclosing the thermofax copy of the check you wrote, but also I am intending to send it back to you, if you could kindly send me a symbolic one (personal) for 1 dollar written to my name (Abdullah Arson) in return. And if it bears the historic date of July 21, I will consider myself the happiest man of the world by having it as an eternal souvenir.

I hope that I was not asking too much. Awaiting your kind reply, I remain Mr. Armstrong, and I salute the owner of the giant step of the mankind in your person. My best and sincerest regards for you and your family.

Yours sincerely,

Abdullah Arsan
Ankara, Turkey

Ankara, the capital of the Republic of Turkey and its second largest city, was the seventeenth stop on (on the fifteenth day of) Apollo 11's forty-five-day, twenty-three-country Giant Step Presidential Goodwill Tour. The astronauts and entourage arrived there (from Belgrade) on October 20 and departed (for Kinshasa, in what was then the Congo) on October 22, 1969. Besides visiting the marketplace, the Apollo 11 astronauts and their wives, with a heavy escort, visited some of Ankara's very old Hittite, Phrygian, Hellenistic, Roman, Byzantine, and Ottoman archaeological sites, including the Temple of Augustus and Rome dating from the first century BC.

Personal reply from Neil

September 30, 1970

Dear Mr. Arsan:

Please accept my apology for the delayed response to your letter of
August 12. I have moved to Washington, and your letter encountered
some delay in forwarding to my present address.

    Enclosed is a duplicate check to the one you are holding. Keep either
one for yourself and cash the other. I will void the check that you keep.

    Enclosed is a photograph which you may find preferable to the check,
in which case you may feel free to destroy the second check.

Sincerely,

Neil A. Armstrong
Deputy Associate Administrator for Aeronautics
Office of Advanced Research and Technology
NASA
Washington, D.C.

"SERIOUSLY NEIL SHE REALLY ADORES YOU"

May 12, 1970

Dear Neil,

Please forgive me for taking the liberty of writing to you, my name is
Johnny Jones. I'm from North Wales been here 5 yrs and like yourself
a father of 4 but, with your help I might be able to pacify my or rather
one of my daughters Charmaine, age 14.

    You Neil are the focal pointing her life she wants to follow your foot
steps, she wants to be an astronaut and no one can change her views
after leaving High School she wants to go to University for Aeronautics
and to take up flying.

I wont' stop her if, she wants to do that. I'll help her in everyway but boy oh boy, you've certainly caused a stir within her.

Your photo's on the table by her bed, she followed your flight to the moon & everything you do, and now: dare anyone or anybody say anything about you (teasing) they are in for trouble as she's rather broad-minded and outspoken in her opinions.

Well this is my Request to you if, you have the time to spare to write a short man to man like letter to her and a photo of yourself with your autograph on it you certainly make a young becoming young lady very very happy indeed.

She's a good kid Neil at the moment with her little sister Yvonne they are making winter coats for the two kittens & Mum sitting in the chair knitting.

Seriously Neil she really adores you, because, of your achievement and as a man & all she's interested is Rockets space & future developments of space programs, if you do consider my Request would you kindly roughly explain the process of becoming an astronaut. Unknowingly to the three of them sitting here I am enclosing some photos of them my wife & my two girls the other big two girls are married with children.

So I'll end now whatever your decision will be in Regards My Request to you, and wishing you & also admire you for your wonderful achieve-ment of human mankind to have your name forever in Human History. God Bless you Neil & your family

Yours Sincerely

Johnny Jones
Stanmore (Sydney)
New South Wales, Australia

P.S. Will you kindly Return the photos to us. Thanks Neil
P.P.S. Our two kittens are named Luna Module & Appollo 11.

"I AM ONE OF THOSE PEOPLE WHO
HAS HEARD ALL ABOUT YOU"

August 13, 1970

Dear Mr Neil Armstrong,

I am Nigel Imrie and I am a pupil in Buca school and I am one of those
people who has heard all about you on radio and news paper and writing
letter at 20 past 8 at night and I am asking you for your photograph
with your autorgraph at the bottom of it.

My home is situated wright on the water front and the passing ships
look as if they were sailing past in our front yard, I am 12 years old
and I am in class 7. On every Friday our class goes to sports we always
play hokey. Our class also has a garden where we plant our vegestables.
Every Friday after we have our lunch our class goes to our garden, in our
garden we plant carrots, spring onions, corn and carrots. Well I must be
going to bed now.

Yours sincearly

Nigel Imrie
Buca Goverment School
Savusavu
Vanua Levu
FIJI

Here I will draw you and your friends when you have land in the sea of
Tranquility.

*At the bottom of the letter is a drawing of the Apollo 11 splashdown with the
lunar module in the water, a helicopter above, and the USS* Hornet *to the side. In
Nigel's handwriting, "FIRST MEN TO LAND ON THE MOON—EDWIN
ALDRIN—MICHAEL COLLINS—NEIL ARMSTRONG first."*

Reply from S. B. Weber

October 2, 1970

Nigel Imrie
c/o Morris Hedstrom Ltd.

Thank you for your very kind letter to Mr. Armstrong and the drawing showing the landing at the Sea of Tranquility. He enjoyed them very much. Unfortunately, because of his heavy schedule of activities for NASA, he was unable to respond personally to your letter.

We are enclosing an autographed picture of Mr. Armstrong as you requested. Best wishes and, again, thank you for writing.

Sincerely yours,

S. B. Weber

"I ENCLOSE TWO POSTAGE COUPONS"

August 16, 1970

Dear Mr. Armstrong,

In July 1969 my family and I were camping in a lovely part of Denmark. One night we got up and walked over the heath to a village to look at the television. It was a long walk for me, but by far not as long as the trip you were on.

That night I saw you on the moon, and later I have seen many pictures of you, but I would be very glad if you would be so kind as to send me a more personal photo and your autograph.

I have postponed this letter till now, as I imagine that you will have better time now than just after the moonlanding to answer such a letter. I enclose a photo of myself to show you and your children what a 13-years old Danish boy looks like.

As it must be terribly expensive for you to answer all the letters you receive, I enclose two postage coupons.

Sincerely yours,

Jakob Bjerre-Madsen
Skovlunde, Denmark

"IT WAS MY GREAT WISH TO SEE YOU"

August 18, 1970

Dear Mr. Armstrong,

Since a long time I am eager to come closer to you—who has touched a new maiden surface. I had to go to many offices in Karachi, to get your adress. Just today I got it from the American centre and am writing you with a hope that you will surely reply it inspite of your busy assignments.

The news that man has conquered the moon was heard with great interest in Pakistan too. Many functions were held in which the speakers praised the moon's conquest. In my school too a programme was held in which a full length film of your moon landing was shown. I wrote an essay about the conquest of moon which was given the first prize.

Mr. Armstrong, it was my great wish to see you and I know you have come to Pakistan but it was my bad luck that you did not come to Karachi. However, I will be pleased if you send me your autographed photo. I hope you will not discourage me.
Wishing you a greater success in the space conquest—and in your life, too

Your Sincerely

Atler Hussain Akbari
Karachi, West Pakistan

"NOW WE HOPE TO GET AN ANSWER FROM YOU TOO"

September 8, 1970

To "The Moon Man"

Dear Mr. Neil Armstrong,

This letter comes from a schoolclass in a little village in far away Sweden. We have of corse been talking about You and even seen films from the moonlanding. We have been wondering if You ever on Your spacetrips have been passing over Scandinavia. But our teacher has told us that Your routes around the world does not touch these parts of the earth. Is that correct?

You must feel honoured Mr Armstrong that we write to You. We have during a couple of years written to many of the most well-known men and women in the world. By this time we have quite a collection of letters and photoes from most kings, queens and presidents in the world—among others from Your own president Richard Nixon.

Now we hope to get an answer from You too—that would be a valuable "number" in our collection. But please do not send only a picture wothout any message at all!

With greetings and hope for at great future for USA in the space.

Class 6, Hallsta School
Eskilstuna, Sweden
Through their teacher Hans Forslund

"SPACE TRAVEL HAS A FASCINATION FOR ME"

September 22, 1970

Dear Neil,

Although I know you very well, you don't know me so to put you in the picture I am a fifteen year old schoolboy who lives in Belfast, Northern

Ireland. This city of ours has been in the news recently but for a reason of which I'm not very proud. On the 21st of July as you well know, you achieved probably the greatest feat of all time—you landed and walked on the moon. On the moon at this moment are the words, "We came in peace for all mankind." I hope people from all over the world will stop and think about this and benefit from what you have achieved.

I am very interested in astronautics and hope to make a career of it working for N.A.S.A. as an engineer. If you could only see my room, and I hope you will one day if you should ever come to Belfast, you would find the whole place covered with every scrap of photographs and data that I could possibly gather on space (and mind you it isn't easy to get this stuff in a city like Belfast). Space travel has a fascination for me. The benefits that can be obtained for the good of man are as limitless as the boundaries of space itself as I'm sure you realize.

Through you I wish to thank—Alan Shepard, John Glenn, Scott Carpenter, Walter Schirra, Gordon Cooper, John Young, James MacDivitt, Charles Conrad, Frank Borman, Thomas Stafford, David Scott, Eugene Cernan, Richard Gordon, Don Eisele, Walter Cunningham, William Anders, Russell Schweickart, and Al Bean. But most of all I praise the crew of the ill-fated Apollo 6 which was a tragic loss to every one. I also want to praise James Lovell, John Swigert and Fred Haize for their tremendous courage and determination when all seemed to be lost during Apollo which instead of a failure turned out to be a magnificent triumph for N.A.S.A., America, and indeed the whole world.

The three remaining people to praise are yourself, Buzz Aldrin and one who deserves special praise for unfortunately being one of the few people in the world not to have been glued to a T.V. on that Monday morning, and yet being nearer to the heroes of the day than anyone else.

I hope you will send me some sort of reply to this letter. If you could it would encourage me more and more towards the exploration of our galaxy when I am older. There will always be an open invitation in our house for you or any of your companions.

Hoping this will reach you.

Carl D. Murray
Belfast, Northern Ireland

September 1970—the month Carl D. Murray wrote his letter to Neil Armstrong—was an especially difficult and tragic month for everyone living in Belfast and all of Northern Ireland. It was an especially violent stretch of time in the Troubles, the sectarian (Protestant vs. Catholic), ethnic, and political conflict that plagued Northern Ireland from the late 1960s to the Good Friday Agreement of 1998.

"I WANT YOU TO BE MY FRIEND FOR EVER"

September 23, 1970

DEAR ARMSTRONG,

Thank you very much for being the first man to step foot on the moon. I have wrote you a letter, but I don't know whether you receive it or not. Please I am begging a free air ticket from you, because I always heard of your name and I always heard of America, but I don't know what is there that is why I am begging a free air ticket so that I can come to you in America. Secondly I want to know your proper adress in America. I am a boy of (15) fifteen years old and I am attending my first year in Latri Kunda School. I want you to be my friend for ever, and I shall always keep on writing to you if I know your number. I am looking forward to your reply.
    Regards to Micheal Collins And Aldrine.

Yours faithfull friend

Kebba Touray
Latri Kunda Junior Secondary School
The Gambia
West Africa

"WE MADE OUR OWN COSTUMES OF ASTRONAUTS"

October 14, 1970

Dear Sir,

I wish you a very healthy and happy 40th birthday. I didn't write to you straight after your great achievement of men walking on the lunar surface for the first time. For I knew lots of people were writing letters of congratulation to you and got the same typewritten letter as everyone else only addressed to a different person. My brother Jonathon (6 years old) and I Garry (10 on August 5th) were so impressed and interested that we made our own costumes of astronauts out of our judo outfit. (In this letter I am enclosing a photograph of us in our costume.) I shall never forget you because our birthdays are on the same date.

Yours sincerely

Jonathon and Gary Glonek
Somerton Park
South Australia, Australia

With the letter came a color snapshot of the two brothers in their "space suits."

"PLEASE CAN I HAVE THE TOP BIT OF THE APOLLO"

October 28, 1970

Dear mr. armestrong

please can I have the top bit of the apollo be cos I want to make a car. I have a sparking plug all ready

thank you

kevin crowley (age 7)
Essex, England

"PLEASE ACCEPT MY MOST ARDENT THANKS AND MY COMPLETE ADMIRATION"

[Translated from French at NASA Headquarters]

October 29, 1970

Dear Mr. Neil Armstrong,

My reason for writing you is because I am very interested in astronomy, but unfortunately I am a girl. I do not see any later opportunity in studying space. That is why, since I admire you a great deal, I hope that you will not find my request preposterous, namely to be kind enough to send my a photograph signed by you.

Thanks in advance. I live in a small village in the South of France. I am 13 years old, in first year high, and come from a working family. Please accept my most ardent thanks and my complete admiration.

Your friend,

Miss Luce Sanchez
Fabrègues, France

"EXCUSE MY VERY BAD LANGUAGE (I'M FRENCH)"

November 6, 1970

Dear Mr. Armstrong,

I allow me to write you because I estimate and maintain your undertaking and your program.

On radio I follow your prowess (?) which are the result of the courage of a whole team.

I have a record which relate the first lunar landing and photos take in books. But I think it isn't enough.

Allow me to ask you something which is worthless for you, but genuine. It will be for me a very beautifully present for Christmas: I think for example of a photo of the moon, or something which has got on the moon and which is worthless for you: a small peace of paper or what you whant.

I tell you the less thing will be to me a great satisfaction and I thank you.

If it isn't possible for you to send me something, I will not be angry. I understand it is perhaps impossible for you. I thank you for all my heart.

Excuse my very bad language (I'm french).

I beg to remain dear Sir yours respectfully.

M. Cognard Michel
Paray-le-Monial, France

N.B. If I can have autographs of Mrs Charles Conrad, Richard Gordon, Alan Bean and Edwin Aldrin, I will be very very happy. Thank you very much.

"THE NIGHT OF THE LANDING, WE DIDN'T GO TO BED AT ALL"

November 28, 1970

Dear Neil Armstrong!

I hope you will have time to read this letter, that is, if it ever reaches you.

My son (than 13 years old) and I followed your flight and landing on the moon, and safe return, from the beginning to the end.

The night of the landing, we didn't go to bed at all, between praying, laughing, hoping, and tears of relieve, we were with you every minute, that is with you, and your comrades, as I am sure you all depended on one another, and oh how I admired your family. At first we were going to write to you at once, but then we said "no, he'll have so much to do, interviews, settling with the family, we'll leave it for a year, and maybe Mr. Armstrong will have time for us."—Because, you see, we have a great request to make to you, and we hope you can fullfill it.

We would like so very much, for you, to write down the words you said when stepping on to the moon, and sign it. Could you please do this. We promisse most sincerely not to show it off, and get you bogged down with requests.
Believe me, you would make my son very happy and proud, and he does deserve a little joy, as he has been a tower of strength to me, since his Dad died two years ago; We wish you all the happyness in the world, and send you our
Sincere regards.

Yours very sincerely

Mrs. Margaret (and Michael) Jennings
Blacon Chester, England

Do your children collect stamps? We could send them some British and German stamps.

Reply from S. B. Weber

December 10, 1970

Dear Mrs. Jennings:

Thank you for your most kind letter to Mr. Armstrong. He is always pleased to hear from friends throughout the world who "lived through those hours" of the lunar landing mission with the Apollo 11 crew. Unfortunately, because of his heavy schedule of activities with NASA, he is unable to personally respond to each letter he receives.

We are enclosing an autographed picture of Mr. Armstrong which we hope you will enjoy.

With best wishes,

Sincerely,

S. B. Weber

"YOU ARE PRACTICALLY THE ONLY PERSON THAT TAKES ME SERIOUSLY"

November 28, 1970

Dear Mr. Armstrong,

I am the girl that drew that sketch of you.

When I received your letter I was so excited that when I went to bed I didn't fall asleep for two hours.

A few days later I also received some photographs of the Apollo 11 mission, including "mission reports" of Apollo's 11, 12 & 13. If you asked NASA to send them, I am eternally grateful. You don't know how much I appreciated them.

You are practically the only person that takes me seriously, and understands how I feel. Whenever I read books on space flights everyone asks me if I want to be an astronaut. What can I say but, yes.

The "Mission Reports" are very interesting. As I was reading them I discovered some things which I didn't find out as I was watching the flight on television. They are very informative.

If you would like, I will send you my version of interplanetary flight. What's strange about it is that it makes sense.

I was very happy when I received your picture. Thank you for everything.

Yours truly,

Kathy Boeskay
Winnipeg, Canada

"I HAD THE HONOR OF WELCOMING YOU
AS A GUEST IN OUR RESTAURANT"

November 30, 1970

Dear Mr. Armstrong,

On 28 July 1970, I had the honor of welcoming you as a guest in our restaurant, Restaurant Wollzeile.

Since we have photographs of all our prominent guests (30 x 40 cm), I would like very much if you could send us a photograph of approximately the same size with a dedication on the front. We hope that you will comply with our request and thank you in advance.

With the hope that we will meet again in Vienna, a Merry Christmas and a Happy New Year to you and yours.

Sincerely,

Rudolf Blumauer
Vienna, Austria

Reply from S. B. Weber

February 1, 1971

Dear Mr. Blumauer:

Thank you for your kind letter of November 30 to Mr. Neil Armstrong. We regret the delay in responding but our mail has been very heavy in recent months, and we have not been able to reply as promptly as we would like.

Mr. Armstrong was not in Vienna on July 28, 1971, and, therefore, could not have visited your restaurant on that date. We have spent some time trying to determine if perhaps one of his colleagues had visited your restaurant and, consequently, been mistaken for Mr. Armstrong. A check with the other American astronauts, however, indicates that none of them were in Vienna on the date specified. We feel certain that someone must have misrepresented himself. Any clarification you might be able to provide would be appreciated.

We do appreciate your interest and extend our best wishes.

With kind regards,

Sincerely,

S. B. Weber

Neil did a lot of traveling in 1971 but none of it involved a stop in Vienna, the capital of Austria. One can only imagine how many times over the years that different men may have represented themselves as an American astronaut—even Neil Armstrong himself. There are other letters in the Armstrong Papers at Purdue like this one.

"EXPECTING MY 'FIRST SOLO' IN THE DAYS TO COME"

December 1, 1970

Dear Mr. Neil Armstrong,

I am just wondering to imagine the impressions of your face when you go through this letter sinse the writter of this letter is a twenty years old boy, who is just learnin about cumulous and cirrus and who try to handle the rudder and elevator of a small aircraft of a plane of the local flying club while the receiver of this letter is . . . . . well. I need not tell you. You know who you are. Why? To tell is short is the last history of aviation which started on December 17th 1903 at Kitty Hawk, North Carolina down to the day when a man landed in the moon, who took off from Cape Kennedy never a man gained so famous a name. Why, you have even replaced Wrights.

I realy take pride that our country honoured you with F.A.I. gold medal for space for the year 1969 during your recent visit to my motherland "for your exceptional performance in space".

I have just started flying though I possessed greater interest to become a pilot for the past 20 years. Only now I am going through clouds where as you got your license even at the age of 17 as I learnt from your biography which I went through many times with everlasting interest.

My instructor told me that he shooked hands with you though you may not be aware of his tugve (?). I definitely believe that one day I will meet you in the long future and will talk to you, and I got great hopes about my future with the only thought that even you would have learnt flying only from the experienced hands of an old instructor. Now not only in the atmosphere & elevation but also in social fame you have been to the higher levels and my sweet and hearty wishes for your prosperous life.

I will be immensly happy if you are kind enough to writte a personal reply to me along with a beautiful photo of yours by the side of a plane. I also want to know, what actually made you to think of becoming a pilot when you are a boy and also your brief opinion of my country that you would have gathered sinse you visited twice.

Expecting my 'first solo' in the days to come with the blessings of people like you for my safe landing!

Though you never speak, as I learnt, I have confidence even a minor thing like this letter will not fail to get your attention, if I am fortunate that this letter reaches your hands.

Counting the days when I will get your reply

Yours "affectionately"

M. Bharathi Ramanathan
Madras, India

"I NAMED HIM NEIL ALDRIN"

December 2, 1970

Dear Mr. Armstrong:

Enclosed please find three photos of my son's first birthday celebrated the past July 27th.

As you can notice on the photos, everything is decorated with the APOLO 11.

The reason why I am sending you these photos is because my little son was born on July 27th 1969, and I promised myself two days before that date, on the 25th, when you reached the moon, that if everything went alright with you my son would be named after you; and so it happened; I named him NEIL ALDRIN, and when I took him to church to be baptized, I told the priest that his Holy name would be MICHAEL.

Since in a very special way I consider you his God Parents, I would apprecciate very much if you would kindly send me your photos (one of each of you) and please dedicate them to my son; I think this will make a very nice Christmas present.

Thanking you for your kind attention to this request and wishing you a very Merry Christmas and a Happy New Year, I remain,

Sincerely yours,

Olga Mayra Fiallo
Mayagüez, Puerto Rico

With her letter Ms. Fiallo included four photographs from her son's birthday party.

"JOIN ME ON A SAFARI IN AFRICA"

December 3, 1970

Dear Mr. Armstrong,

I must confess that I am COMPLETELY confused as how to begin this letter to a man as important and World re-known as yourself.

This letter is merely an invitation to you to join me on a safari in Africa should you be able to do so, and at any time you care to do so.

I would like to emphasis before I go any further, that I am not extending this invitation for personal gain from the publicity that may come out it should you accept. In fact, should you so wish, I would do my utmost to make it a private and secret safari.

I am at present, visiting the United States, and I will shortly be returning to Rhodesia. Should you decide to accept this invitation, I would appreciate an early reply.

I would also like you to know that this is a genuine invitation, and I am not a "crack-pot" looking for publicity.

I have enclosed a business card of an associate at whose address I would appreciate you sending your reply.

Look forward to hearing from you,

Yours faithfully,

Richard Whyt
African Camping Safaris
Salisbury, Rhodesia

Personal reply from Neil

January 22, 1971

Mr. Richard Whyte
c/o Mr. Joe Kulis
Kastaway Kulis Taxidermy Studios
Bedford, Ohio

Dear Mr. Whyte:

Thank you for your kind letter of December 3, and please accept my apologies for the delayed response. I managed to take off a couple weeks during the holidays and, therefore, got behind in the correspondence department.

Your invitation to join you on a safari is most appealing; I'm sure it would be a novel experience. It was particularly thoughtful of you to offer anonymity to me in the event I should be able to accept. Unfortunately, my heavy schedule of activities with NASA precludes my scheduling personal pleasures very far in advance.

I will keep your invitation in mind, however; and, should the opportunity for some time of my very own while in the vicinity of Rhodesia present itself, I will get in touch with you. I don't mean to sound optimistic, however, as my new duties have been rather demanding the past few months; and there is no indication that the situation will change any time in the near future.

Best wishes for a pleasant and rewarding 1971.

Sincerely,

Neil A. Armstrong
Deputy Associate Administrator for Aeronautics
Office of Advanced Research and Technology
NASA Headquarters
Washington, D.C.

"WHEN YOU LANDED, WE HAD NATIVE DANCING, AND WE ALL CELEBRATED"

December 17, 1970

Dear Mr. Armstrong,

First I would like to introduce myself. I am an 18 year old, and am from a country that you may not have heard much about, but what you may have seen when you were up in Space. It is Papua-New Guinea.

I cannot describe to you the night you and the other men were landing on the moon. All of us waited on the beaches and in our villages listening to the radios. When you landed, we had native dancing, and we all celebrated. Then some of the small children said they could "see you" walking around. Do you think that you would be able to send me a photograph of Mr. Collins, Mr. Aldrin and yourself? Also would you please sign your names on it for me? When I receive it, I would like to show all the boys in our village. They admire you all as of your courage, as it was something others had not done before and you men faced it up.

I hope that I may hear from you and it would make me very proud. Would you also be able to say hello to your familiy for me, and I wish them a Merry Xmas.

I hope that oneday you may all have a tour here to Rabaul, to visit us all.

Wishing you all a Merry Xmas and A Happy New Year. Please write oneday before you go to the moon again.

Yours sincerely,

John C. Goad (Jnr.)
Rabaul, New Britain Island
Papua New Guinea

Reply from S. B. Weber

January 27, 1971

Dear Mr. Goad:

Thank you for your kind letter to Mr. Neil Armstrong. He is always pleased to hear from friends throughout the world. Unfortunately, because of his heavy schedule of activities with NASA, Mr. Armstrong is unable to personally respond to each of the many letters he receives.

We are enclosing a crew picture with Mr. Armstrong's autograph. However, we are unable to furnish the autographs of the other crew members as the three are no longer a team and are located at separate locations. You may send the picture to them at the following addresses, and I'm sure they will be glad to autograph it also:

The Honorable Michael Collins
Assistant Secretary for Public Affairs
Department of State
Washington, D.C. 20520
U.S.A.

Col. Edwin E. Aldrin, Code CB
Astronaut
NASA Manned Spacecraft Center
Houston, Texas 77058
U.S.A.

Best wishes and, again, thank you for your interest.

Sincerely,

S. B. Weber

"I KISS YOU!"

December 24, 1970

Dear, nice, beautiful spaceman Neil!

I should be very happy if you can send me one big photo 18 x 24 of
yours and another little photo 7 x 10 to keep in my bill fold (please,
with your dear dedication)
    You are the most wonderful and pretty, nice spaceman of the
United States!
    I wish you happiness and to carry many days
    Happy new Year for you my love!
    I Kiss you

your young FAN

Linde Montello
Bolzano, Italy

"WILL YOU BE FRIENDS WITH ME BY MAIL"

January 3, 1971

Dear Mr. Armstrong.

My name is André and I am 11 years old. I adore space and everything which is conected with it. I also am writing a book about space, I wrote already 700 pages. When I am grownup I will be an astronaut like you Sir. Since you are the first man, wich landed an the moon, I would like to make friends with you. I would like to meet you in person but that's not possible since I live in Brasil and you in the U.S.A. So I ask you, please will you be friends with me by mail. I would be so happy to have you as a friend.

Many thanks in advance for your attention:

Yours

André von Hebra
São Paulo, Brazil

"MILLIONS OF PEOPLE WERE ACCOMPANYING YOU"

January 6, 1971

My dear great hero Armstrong:

It is a wonderful occasion indeed to write this salutation to the world's greatest hero, the first person to step on the surface of the moon, scoring a world event that has never happened before since the beginning of history. Man, that great explorer, and the first and only penetrator of the mysteries of the universe and space, has been able through his bright mind and brilliant thoughts to explore the planets and stars, and you are the very first person and hero to explore and penetrate those mysteries.

I admire your great heroism and courage, and, believe me, you were not alone in space. Millions of people were accompanying you with their hearts and souls while you were putting your foot on the ground of that white planet. The moment I saw you getting down from the module and stepping on the surface of the moon was the happiest and most pleasant moment of my life. My eyes had never seen such an enjoyable view.

I, a painter artist from Lebanon, wish to express my admiration of your accomplishment not only through the paintings I have made but also through the few lines of this letter which I am sending you with my heartfelt and most sincere regards and admiration.

I will be thankful if you could send me your autograph or any other souvenir from you.

Your artist admirer

George Haddad
Beirut, Lebanon

"SIR, I HAVE HEARD OF YOUR KINDNESS"

January 26, 1971

Dear Sir,

I have the respectful honour in writing you this my humble letter. But before I say a word I have to ask of you present condition of health. Sir, I have been hearing much about you and how I have been hearing people talking about you and your Company went on space to the moon and I am very happy to hear about the news. So please if you will be kindly send me one or two of your best photograph. Sir, I have heard of your kindness that is why I am writing you to send me your photograph, so that anytime I sees it I will be remembering you. So I will be very glad if you send it to me in time. I stop here greeting to you and your company.

Yours Truthly

Yahaya Issah
c/o Issah Yahaya Kpamahin-aa

Tamale N/R [Northern Region]
Ghana

"WHY IT IS SAID THAT A THING CAN NEVER HAVE A SPEED SAME AS LIGHT?"

July 1, 1971

Mr. Neil Armstrong:

Hello. I am an Iranian girl 16 years old. I am interested in the asterogical matters. So I have collected photoes and articles from this matter. I want to ask you to have corrosponding with you about these matters.

Of course you may be so busy that haven't time to answer me, but I ask you to answer Just this letter of mine, and make me very happy.

Well, sometimes I face to something that Nobody can answer me, but you. These are my questions:

1. Why it is said that a thing can never have a speed same as light?
2. Why does the conection of a rocket off when it arrives in the atmosphere of the earth?
3. Why it is said if a rocket for example has a speed same as light if it spends two years in reaching to a planet when the airmen reach to that planet thousands years have past from the life of the earth?
4. What is the meaning of proof of material article?

Thank you

Yours affectionately:

Mina Honarkhah
Teheran, Iran

The Apollo 11 crew visited Tehran for three days on its Giant Step tour, arriving there on October 24 and departing on October 26, 1969. While there, Armstrong, Aldrin, and Collins, along with their wives, had a private dinner with Mohammad Reza Pahlavi, the shah of Iran, and his family. (Pahlavi was Iran's last Shah, overthrown by the Iranian Revolution on February 11, 1979.) A photograph and film were taken of Neil showing a rocket model to the empress consort of Iran, Farah Pahlavi, and two of her children. The shah presented all three astronauts with awards and each of their wives also received a gift.

"WE WOULD LIKE TO OBTAIN THE CITIZENSHIP OF THE MOON"

July 9, 1971

Dear Mr. Armstrong,

About four weeks ago we wrote to the American Embassy there inquiring about citizenship of the moon. We received a letter from the Secretary saying that they had no information on how to go about getting the citizenship of the moon, but they very kindly sent us a book on space, so we decided to write to you to see if you could help us.

We would like to obtain the citizenship of the moon for our younger brother, because his name is TRANQUILLO and he was eleven on July 20th 1969 when Apollo 11 landed on the Sea of Tranquillity. We read on a newspaper that a man named TRANQUILLO obtained the citizenship of the moon.

Well, as you see, there are three coincidences linking my brother with the landing of the Americans on the moon.

We would be most grateful to you if you could send us the Citizenship of the moon, if it is possible, as it would be a very nice and original present for his next birthday. It does not matter if it cannot be obtained, we shall still be grateful.

We thank you very much, Mr. Armstrong.

We remain your truly

Sandra and Flavio Sidoh
Parma, Italy

"MY FATHER AND I HUGGED EACH OTHER IN DEEP EMOTION"

July 18, 1971

Dear Mr. Armstrong,

I am taking the liberty of addressing you, I do so as a man, as a marine and as an admirer of the great deed achieved by you and your comrades in adventure—the conquest of the moon.

When the great deed occurred, I was home with my parents taking advantage of several days leave and, since television had not yet arrived, we only listened on the radio to all the details of that historic event. At that moment my father was the most enthusiastic because everything was a success. Unquestionably I as a keen admirer of the north american space programs, listened and hoped with deep faith that everything would turn out with no problem. That is the way it happened! My father and I hugged each other in deep emotion because a "north american" had been the first human being able to set foot on the lunar soil. This was the conquest of the moon, the wonderful star of the night which I admire. This deed also fulfilled the visionary words of Jules Verne, the peaceful wishes of Wernher von Braun of reaching it someday, and the precise vision of the unfortunate President John F. Kennedy.

Then came the happy return to earth, quarantine, return home, praise and a return to the world including the country named Chile. A little of the enthusiasm for reaching the moon was calmed, but the achievement was still crowned with real success; I was interested in listening to radio programs where there was always talk of the thrilling subject of the star and, more than that, I dedicated myself to the purchase of books which gave me a more objective and clear vision of what such a great event had been. But, the best thing was seeing the colored film where I seemed to go wondering through the vastness of the universe.

When I wrote to my friends that year I always mentioned the deed accomplished by you and your colleagues. On one of these occasions I wrote thus:

When on 31 December last the hands of the clocks marked 2400 hours, the end of one year and the beginning of another, we left behind a transcendant decade in the history of man and all humanity. Another decade, the decade of the 70's, had also begun with it. It is also uncertain, but from future and unforeseeable eventualities throughout time there will be similar things in their unspeakable significance in the world of modern cosmology and in deed for the world of those who do not reason. For the genius of man nothing seems to escape the deep and marvelous field of science. I am lucky to have been born in a generation of men of such intellect who have been capable of transporting us beyond our imagination and to the vastness of the universe; we had been optical witnesses of the conquest of the moon by them.

Since the moon is second only to the sun as the most attention-getting star for men, it was evident that someday they should try to reach it . . . and they did! Exactly in 1969 was the year that man first walked upon that pale but romantic star of night. With this unimaginable event on 20 July began year 1 of the space era. In this way were opened the routes toward interplanetary flights, towards space stations and toward the conquest of the other stars located beyond distance and silence and sinking into the unknown. Later science will put other men to the test who will succeed in reaching that goal with as much courage and intelligence as the first.

But for that 20 July 1969 which will end another year in a few days, as you said, "It was a small step for man but a great jump for mankind." For me, you and your companions it can be defined in this way: two heros immortalized in the field of science. Two men whose looks were lost in the indescribable image of the star will remain forever floating in the immensity of the moon and everything that was seen on that day will remain forever engraved on the retinas of the astronaut's eager eyes.

Even at the beginning of time there were many men of science who, with their precarious means, were interested in the stars which surrounded them. But a great deal of time had to run out before other men of other generations were to become interested in the attractive subject of the milky way.

From the dawn of what today is cosmology, they were born for science: Pythagoras, Democritus, Philolaus, Aristarcos of Samos, Hiparcus and others. All of them handed down valuable data, discoveries and inventions to the next generation of men called geniuses and even "fools" by their own countrymen. Beginning with the tenth century and running to 1548 in which it appeared: Copernicus, the great Galileo Galilei, Tycho Brahe, Kepler—who discovered the famous tables of Kepler—and with this the introduction of mechanics into astronomy, and thanks to the genius of Newton even universal gravitation was formulated by Kepler. Thus was founded and introduced celestial mechanics, a branch of astronomy. Beginning from that epoch, the science of the stars entered its modern phase. Since that time work has been increasing, both in the material subject and in physics. With the previous data, work, discoveries and inventions, scientifically interesting, notice was taken of the unsuspected evolution and development in which is found the obligatory subject of our day: the universe which surrounds us and the possibility of our current and future generations to explore and become acquainted with it all. The reality in all this is that we are living and experiencing the most amazing moments of the history of humanity . . . the exploration and conquest of interplanetary space. This is the reason for my effort and interest in writing to exalt the deed of the century and the brave space personalities, the work of genius of Wernher von Braun, the father of modern rocketry.

I know that you are familiar with this subject in a better form; I only wished to make known my ideas and thoughts on what all this meant to me with a brief summary of the subject and its history.

The main reason for this letter is to ask you to please send me a photograph taken from the lunar module sailing above the lunar space with the earth appearing in the background on the horizon. You can send it in the safest way possible or by Navy. My address is on the cover. Again my admiration for you and your comrades and for a desire for future success in the ambitious Apollo program which has become so famous throughout the world.

I would also like to send you a colored photograph of some section of this long and narrow land beneath the southern sky which is called Chile.

Sr. Roberto Perez
Valparaiso, Chile

Personal reply from Neil

October 4, 1971

Dear Sr. Perez:

Thank you for your very kind letter. The tributes paid the Apollo Program and the flight of Apollo 11 have been most gratifying and I appreciate your taking the time to add yours to the many others that my colleagues and I have received.

We hope your interest and enthusiastic support will continue as our activities in space exploration progress.

Sincerely,

Neil A. Armstrong
Deputy Associate Administrator for Aeronautics
Office of Advanced Research and Technology
NASA
Washington, D.C.

"WHAT SORT OF MAN ARE YOU"

July 26, 1971

Mister Armstrong,

First, I congratulate you for your birthday, then I recongratulate you for your courage that you had had when your mission on the moon.

I am 15 years old, I am french and if I dare write to you, that is for the reason that I am one of yours admiraters. For a long time I hesitate to write this letter and today, I take my pencil and my courage for to ask you some questions.

In school (I was in 2 Technic) with the teachers, we spoke often of the space, and always with the sciences, physics and technology of construction teachers, we had made a bundle of papers but the times passed and we had not finished it, and now it rest many empty place in the book, then this is on this moment that you interfere.

There are:

1. What sort of man are you and what sort of things like you?
2. What impression had you feel when you had put yous feet on the moon?
3. To support the atmosphere who is on the moon what is the material whose are made the instruments who are rested on the astre? of steel? Mild, half-hard, hard steel? Of Alpase, of duralumin?
4. The pieces are casted or are labored with fashioning?
5. The newspapers said that the moon is a portion of the world who was before at the Pacific's place, is it real?
6. On the moon, what sort of mines can we found?

Please can you send me a photography of you and your family and another lunar's stones. I thank you very much for the derangment that I made to you but my bundle of papers will be (if you received my letter) grace to you full. I once more thank you.

Yours very truly, kisses to the littles children.

Sincerly

Dan Gosse
Casablanca, Morocco

"DEAR NIL ARMSTROG!!!"

August 1, 1971

Dear Nil ARMSTROG!!!

I am an Israely-boy. I heard a bout your journey to the moon with the spacship.

I fllow after the spacship with radio and T.V., and newspapers.

When you and Oldreen (forgive me if i wrote the nam bad) go out from the "Eagle" I was very happy and when you and your friend come to the Ehrth I was very happy

I don't have time to writ more but I can say: It was a completely journey.

And now I want (if you can), send me your pictur and writ my nam and your signature on the pictur

Thank you very much!

and with many thanks!

SHOKRON Shalom (13 years old)
Dimona, Israel

"WHAT DO YOU CONSIDER YOUR MOST THRILLING MOMENTS IN SPACE?"

August 3, 1971

Dear Mr. Armstrong:

I am writing to wish you a very Happy Birthday.

Last December, Mr. S. B. Weber sent me a photo of you which you had personally autographed for me. You dated the photograph Nov. 30, 1970 and he sent it Dec. 2, 1970. I want to thank you for personally autographing a photo for me. Until that time I had tried unsuccessfully to obtain your autograph for two years, well before Apollo 11. I am glad now to have it.

How did you get to be an astronaut? How did you hear about astronaut selections? When, where and by who were you told you had been selected as an astronaut? What was there about NASA and space travel that interested you?

Were there times either on Gemini 8 or Apollo 11 when it seemed you were in simulators rather than in space?

I have heard that some astronauts take their own cameras into space. Did you ever take any cameras of your own into space?

What did you find more exciting or nerve wracking—Gemini 8 and Agena tumbling in space or the final few minutes before lunar touchdown or lift-off?

I would like very much to obtain any materials from Gemini 8 if you could get a hold of any photos, press releases, and other written materials would be very nice to have.

What was the exact problem on Gemini 8 and why did it happen? I have some idea that it was a stuck thruster in the ON position but I'm not sure if that is correct or not.

I began my interest in NASA and manned space flight when I watched the flight of Gemini 3 on television. Since then I have become more and more interested in NASA and I want very much to work there—probably in the photography department. I hope someday to fufil that goal.

Did you photograph your *first* step onto the lunar surface? If so, I would like very much to obtain a copy of that photograph.

What do you consider your best *and* worst moments of being an astronaut?

What do you consider your most thrilling moments in space? And what was your worst?

Knowing that in all likelyhood you would never set foot on the Moon again, did you leave or bring back anything from Tranquility Base?

What did you find your best mode of travel on the Moon? Did you ever fall down at Tranquility Base?

I would welcome any information—photos and written materials—on the flight of Apollo 11. Such items as flight plans, press releases, press conference transcripts, press kits, lunar trajectory notes and so on would be extremely nice to have.

Who selected the names "Eagle" and "Columbia?" Did the CMP choose the name for the CM, and the LMP for the LM or did all three of you decide on the names? How long before the flight were they choosen?

I would like to have some information on the first words you spoke after putting your left foot on the surface of the Moon. No doubt you thought about what you would say before and during your flight to the Moon.

However, as I have listened to your words it seems to me that your final version of what you would say came as you actually said them. You paused, then said "That's one small step for man," paused again, then said "One giant leap for mankind." Did you actually think of what you were going to say then or when actually did you decide on what you would say?

I suppose it is because I think so highly of NASA and its personnel that it bothers me when people criticize it. Naturally it is alright to criticize but people should attempt to find out more about it and the many achievements for the world as a whole that have come about from the NASA program. It also "bugs" me when people, including written NASA material, do not quote your first words exactly. Would you confirm for me, that your first words on the Moon were: "That's One Small Step For Man, One Giant Leap for Mankind." Time and again I have heard and read "for *a* man." It not only is inaccurate but takes away the strength of your statement. Does it ever bother you that you are not always quoted correctly?

I think your first words were excellent, but have you ever wished you had said something different?

Do you now live in Washington? Why did you not stay in Houston? At the NASA Headquarters in Washington, what is your title and what are some of your duties? I would like to obtain transcripts of speeches that you have made. All if possible.

If you could obtain any photos that have been screened and failed from your flight or others, I would like very much to obtain them. I understand that one copy is made from each negative. The photo is studied and either passed or failed. Those failed are either given away or disposed of in some other manner. Is this correct? If these photos can be obtained on Kodak paper that would be fantastic.

I have written you a very long letter and I know you are very busy. I would appreciate your replying to my letter but when you do it, do it at your convenience. I don't care if it takes six months or longer but I sure would appreciate your replying to my letter. If you prefer, answer by handwriting rather than typed. It is always nicer to have something hand written but it also takes longer to do.

Two final and very important requests.

I would appreciate it very much if I could obtain from you an Apollo 11 flight patch. On July 16, I was extremely elated to receive an Apollo 15 patch from Jim Irwin. I couldn't believe it! I have one other patch, that from Apollo 10. After writing to NASA for the past 2½–3 years, I know that flight patches are not normally given out to those people who are not NASA personnel. I am extremely proud to have obtained these patches and am hoping I can add yours to it.

I guess you probably took a couple of hundred photos during the mission of Apollo 11. I would appreciate it if you would choose your favorite photos from Apollo 11 and put on that photo your first words as you stepped onto the surface of the Moon. Thank you so very much.

I hope you have a very Happy Birthday and all the best. Enclosed you will find some items on my province's Centennial.

British Columbia entered Confederation on July 20, 1871. July 20 is a very popular date when it comes to history! I hope you and your family will find some enjoyment from these items.

If this arrives late I apologize. I have been busy watching the flight of Apollo 15. That Gemini 8 crew sure made up for it!

Thank you very much for taking the time and the trouble to read and answer my letter. I sincerely appreciate your efforts.

Yours sincerely,

Trudy Hopkins (Miss)
Penticton, British Columbia
Canada

Reply from S. B. Weber to Miss Hopkins's letter received in November 1970

December 2, 1970

Dear Trudy:

Thank you for your recent letter to Mr. Armstrong. He is always pleased to hear from young people with a keen interest in our space program. Unfortunately, because of his heavy schedule of activities with NASA, he is unable to personally respond to each letter he receives.

We are enclosing a "Log of Apollo 11," which contains the answers to your many questions and some of the photographs taken during the historic lunar landing. Should you wish to order additional pictures, you may order those of your choice from the U.S. Government Printing

Office at nominal cost as indicated in the attached brochure entitled "Man and Space."

Mr. Armstrong has autographed a picture for you which we are also including.

Best wishes and, again, thank you for writing.

Sincerely,

S. B. Weber

"TAKE JAMES BURKE FROM BBC-TV ALONG WITH YOU"

August 11, 1971

Dear Neil

I feel I must write to congratulate you and your colleagues on the fabulous success of Apollo 15. You will probably receive many such letters, but over the last few years I have followed so many of your unbelievable flights that I feel I must write to at least thank you all for bringing a great deal of excitement into my dull little life!

It was the Apollo 8 flight that really got me "hooked." To be quite honest, if it hadn't been Christmas, & being Christmas over here, there is nothing else to do but watch TV, I would probably never have got better. Since then I have been plastered to the TV for each of the succeeding flights. Each brought its own particular problem for me. Trying to figure out if I'd have to get up in the middle of the night, playing sick in order to get off work to watch moon walks, & making a mad dash to get home in time to see launches, splashdowns, or whatever. For the lift off from the moon last week I think I created a world record on my own. It's normally a good 20 minutes walk home at night, last morning I made it in just over 10 minutes. It's uphill all the way too, I hardly had strength to switch on the TV. I sort of collapsed in a heap on the couch. I had been thinking of trying to wargle the day off work, but as I'd been taken suddenly 'ill' the previous Monday in order to watch the launch from Cape Kennedy I didn't think I'd better. It's perhaps as well that there are only two more Apollo flights. I am fast running out of excuses for taking days off, & sooner or later my boss is going to come to the conclusion that it's strange that I'm ill every time there's a moon landing. I don't know what I shall turn my interest to when all Apollo flights are over though. As I said, they bring a little excitement into my dull little life. Believe me my friends & I at work are planning our annual trip to Liverpool with as much care & detail as you must plan your moon landings. We go every year just for a day, but although your moon landings usually go as planned, our day trip to Liverpool always goes haywire the minute we set off. There's usually four hours of torture on the boat going & returning on the toughest sea possible & it usually pours with rain too.

I know you will think this a heck of a request, but over the last two or three years, I have grown rather fond of you fabulous men. Could you possibly send me a photograph of each of the Apollo crews? It must sound very greedy but I think you are all so great. You're all so cool & calm about it all.

Oh, one small favour, would you please make Apollo 16 large enough to take four men & then you can take James Burke from BBC-TV along with you. He is the only thing that spoils the moon landings, he's so busy nattering away it's impossible to hear what's being said between the spacecraft & Houston. He nearly drives me insane, so if you could possibly take him with you next time & leave him there, I would be forever in your debt!

Yours sincerely,

Denise Kay
Isle of Man, British Isles

"BRING YOUR CLOTHES TOO"

[Unspecified date, September 1971]

I would like some comppany.
   I want you to come over to My Place. For over night. Bring your clothes too. Becuase I want to Lean about space.

Love

Brent
Brent Buhler
Calgary, Alta.
Canada

Reply from Geneva Barnes

September 28, 1971

Dear Brent:

Mr. Armstrong has asked me to thank you for your note inviting him to visit you. He receives many similar invitations and I'm sure you understand that he is unable to accept because of the demands on his time.
   We are enclosing an autographed picture of Mr. Armstrong which we hope you will enjoy.

Sincerely,

(Mrs.) Geneva Barnes
Secretary to Mr. Armstrong

"DIDN'T YOU THINK TO STOP WITH THIS MOON-FLIGHT"

November 8, 1971

Dear Mr. Neil Armstrong,

Here's a letter from a boy, who interests Space Travel very much. Excuse me, my name is Karel Mras and I live in Heerlen.

I'll ask you some questions. I hope you are not angry, because I wrote you, without asking before.

It a only a few questions, because I now you have not much time. First question:

1. Mr. Neil Armstrong, where do you really live.
2. How do you feel as first man on the moon.
3. Neil, In 1966, you wernt for the first time in Space. What was the real reason for a landing in the ocean unexpected.
4. Could you explain in a few words, how it is to be in Space with two men during a few days.
5. The name "Eagle" who found it out. Micheal Collins, Edwin Aldrin, or yourself.
6. Your flight was very interessant and now I will ask you, how do You really think about Space-Travel.
7. Did the Landing in the ocean, came hard or soft.
8. Was it your opinion to be a flyer.
9. In May 1968 you risked your life by training. Didn't you think to stop with this moon-flight.
10. Mr. Neil Armstrong, what are you going to do in the future?

Mr. Neil Armstrong, I hope you can read and understand what I write. My english isn't so good as you may be expect. But I hope I get an answer on the questions, Mr. Armstrong.

Thank you in advance

Karel Mras
Heerlen, Holland

"CERTAINLY, YOU'RE ASTONISHED GETTING THIS LETTER"

December 12, 1971

Dear Mr. Armstrong,

Certainly, you're astonished getting this letter in your hands. Where did it come from? Well, from Zaïre, and written by a young student of T.T.C. I'm studying at Luma, in the North East of our country. My language is french but I'll try to explain me in English.

I wrote you this letter to ask you, if possible, will you explain me the meaning this initial of your company: N.A.S.A.?

I'm very excited to learn it because on all your clothes, engines and pictures, and I'm reading it without knowing.

Will you help me, Mr. Armstrong?

With all my school-fellows, hoping for an answer, I send you our best greetings.

Will remember us to your family.

Yours faithfully,

Jean Pierre Abdala
Ana, Uganda

Reply from Geneva Barnes, Neil's secretary

June 30, 1972

Dear Sir:

Mr. Armstrong has asked me to thank you for your letter and the kind sentiments expressed therein. He is always pleased to hear from those who are interested in the space program and would like to reply personally to all. However, I'm sure you realize this is not possible because of the demands on his time.

I apologize for the delay in answering, but our volume of mail is very heavy and we are not able to reply as promptly as we would like.

To answer your specific question: the letters *NASA* stand for National Aeronautics and Space Administration; this is the agency of the United States Government charged with carrying out the stated policy of the United States that "activities in space should be devoted to peaceful purposes for the benefit of all mankind." There is enclosed a brochure on NASA which explains its functions and responsibilities. We are also enclosing an autographed picture of Mr. Armstrong and some information which we hope you will enjoy and which will answer many of your questions.

Best wishes for a rewarding school year.

Sincerely,

(Mrs.) Geneva Barnes
Secretary to Mr. Armstrong

"WE ARE AWAITING THE BLAST OFF TOMORROW OF APOLLO 16"

April 15, 1972

Dear Mr. Armstrong,

I feel I must write to thank you very, very much for sending me, through my mother, a signed photograph of yourself. I have followed the American space programme for many years now but, without doubt, the most thrilling experience I have ever had was the Apollo 11 moon landing. However, each mission has had its own character and thrills and credit is due to all participants in the U. S. space effort. We in England have been provided with excellent T.V. and press coverage of all the major events in the space "callender."

I have, over the years, bought countless magazines, newspapers and books on astronomy, rocketry, the moon and of course the U. S. space programme. I am particularly pleased & proud to be named Neil in common with yourself, the *first* man on the Moon. My ambition is one day to visit the Kennedy Space Centre and see for myself a manned space launch. My wish is that the U. S. Government will be benevolent and generous to the Space endeavor in order that further projects may be undertaken. It must be very worrying to think that money is getting short or that the efforts of so many people may go unrewarded in terms of future missions and developments.—Mankind has a great deal to thank you *personally* for in proving that man can leave his own planet and venture into space and the unknown. Your great courage and personality will be forever cherished by both the present and future generations.

At the time of writing we are awaiting the blast off tomorrow of Apollo 16 and I would like to take the opportunity of wishing Astronauts Young, Duke and Mattingly God's speed for a successful mission and a safe return to Mother Earth.

The future of mankind I believe lies in Man's conquering space and without doubt future space missions will capture the imagination of millions of people the world over—my dearest wish is that the people of the earth will, through the experience of manned space exploration, realise that we are but a small speck in the universe and that the troubles of the world may not be so great and insoluble as are feared at present.

Thank you once again for your personal interest and kindness and I hope that one day, somehow I may meet you in person.

Yours admiringly,

Neil S. Dearnley
Leeds
Yorkshire
England

"YOUR EXPLOIT WAS EXTRAORDINARY AND WONDERFUL"

December 16, 1972

Dear Mr. Armstrong,

I have the pleasure to write to you. I haven't been early since Apollo XI. My name is Françoise Contant. I'm a french girl. I'm 15 years old.

First, I want to congratulate you very much. Your exploit was extraordinary and wonderful. It made a lot of hearts beat. But now, the enthusiasm disapears. Why? Because, it's the routine! It isn't a good reason I think. The poor last were almost alone on the moon, while you were followed by thousands of burning eyes. For looking at them, I was in a sultry room of a boarding-school and I envied them. Gazing on the impression of one of the men's foot, slight and at the same time boundless human memory, I was more than moved.

Allow me to ask your impression when you were on the moon, and after when you have seen other men on it.

And can you send me your photograph with Collins and Aldrin.

Yours respectfully,

Françoise Contant
Fours, France

Happy Christmas for you and your family.

"THE DISTILLERY HAS DONE YOU GREAT JUSTICE"

May 11, 1973

Dear Mr. Armstrong,

On a recent trip through a local market place here in Tehran I came upon an item which you may find amusing should you not have known of it before.

It was a genuine bottle of Neil Armstrong Brand vodka. Enclosed is the label I soaked off the bottle after personally soaking up its contents. Feel proud, Mr. Armstrong, the distillery has done you great justice in the manufacture of this fine spirit.

It was one of my grandest moments to sit in my own living room and vicariously relive your first historic steps on the Lunar Surface. Actually, I think it was my first historic steps on my own Lunar Coffee Table also (simulated weightlessness being achieved on my Lunar Sofa).

Yours Truly,

Henry Giamella
Iran Aircraft Industries
Tehran, Iran

5
FROM ALL AMERICA

As much as Neil Armstrong meant to the world, he meant even more to America. He was, after all, a genuinely all-American boy, who grew up in "Smallville, U.S.A.," deep in the heart of Middle America, in various small towns within the State of Ohio, to humble parents, surrounded by small farms and vast cornfields. As much as America's comic-book hero of the same era, "Superman," promoted "Truth, Justice, and the American Way," so, too, did Neil, as a dedicated Boy Scout (who reached Eagle Scout status), baritone horn player in his high school and university marching bands, naval aviator who fought for his country in war, experimental research pilot test flying the most advanced experimental aircraft of the age, and intrepid astronaut. That it was an archetypical American who commanded the first Moon landing, and first stepped out onto another world, seems, indeed, to be the logical culmination of a fundamental essential American destiny, the culmination of generations of adventurous ambitions, pioneering spirits, and exceptionalism. That Neil's ultimate achievement—becoming "First Man"—was truly not anything predestined singularly for Neil (or anyone else), but was rather a destination in an improbable journey through a labyrinth of historically and biographically contingent circumstances, was a fact Americans largely misunderstood, ignored, or, most of all, preferred to interpret in a more glorious and heroic way.

What Armstrong came to mean to America and Americans can be seen and more clearly interpreted through the prism of the many thousands of letters he received in the years following his Apollo 11 mission. That clarification comes from analyzing not just the letters he received from the country's young people, which came in droves, but also those cards and letters sent to him by men and women of his own age and older (some much older), many of whom saw in Neil the realization of their own dreams.

The items of correspondence in this chapter have been carefully selected to reflect the rich diversity of materials Neil received. Many hundreds more letters in the Armstrong Papers could easily have been used productively in this chapter. Following several of the letters is a reply from "Neil," most often from his secretary or assistant at the time (at NASA Headquarters, at the University of Cincinnati, or personally employed by Neil), but in some instances the reply comes from Neil himself. Even when written by an assistant, a reply still very much reflected what Neil wanted to be said. All of his assistants over the years worked very closely with him and knew what he wanted (and what he didn't)—Neil gave rather precise instructions. For letters containing one or more frequently asked questions or covering subject areas commonly asked about, the assistant would work from a sample letter, tailoring it as necessary.

Most of the letters Neil received were very friendly and cordial, and several in this chapter exhibit elements of true hero-worship. Readers will find a few examples of not-so-friendly letters as well. These came to Neil more often than one might imagine, especially when an earlier request had not been replied to in a way that fully satisfied the letter writer. Such letters show how difficult it was to be First Man because so many people expected such great and immediate positive replies to each of their requests, some of which could be extraordinarily greedy and selfish.

"HOW YOU DOING, DOUG!"

August 8, 1969

Dear Neil Armstrong,

Congratulations on your great, historic flight. I am a 14 year old boy who wishes he could have been right up there with you.

I'm writing you this letter to ask you for a small favor. It would be a very, very, great thrill for me if you would write a little message to me and autograph it. It doesn't have to be much, just so I knew it came from you. "How you doing, Doug!" would be fine. It would mean an awful lot to me, and I could always use it in school with great pride. To make it easy for you, I will enclose a self-addressed envelope so you can just put it in there and mail it. I hope you will consider it, you don't know how it would mean to me!

Thank you so much,

Douglas R. Conway
Roanoke, Virginia

From the boy's mother

August 10, 1969

Dear Neil Armstrong:

The attached letter of sincerest "hero worship" is from my fourteen-year-old son, Douglas, who is suffering from a terminal malignancy. He does not know the gravity of his health problem, and I *do not* want him to know; however, I thought that if *you* knew how little time he has left on this earth and that it is within your power to give him the biggest moment of happiness in his short life, perhaps you would be able to find a few moments in your overcrowded schedule to write a few words to him.

As his mother, I should be eternally grateful. As I see the hours of his life run out, I would give him the moon, if I could. YOU are as near as I can come to giving him his greatest heart's desire; thus, I humbly *beg* you to help me to give him this special moment of happiness and fulfillment. It would come as a sort of glorious climax at this time.

Most earnestly and sincerely yours,

Nancy M. Conway
(Mrs. Howard V. Conway)
Senior Clerk, Postal Inspection Service

Personal reply from Neil

August 28, 1969

Dear Doug:

Recently I received a letter from you which I enjoyed very much. I am always delighted to hear from our young people—especially those with your kind of interest. I am enclosing an autographed picture of our crew which I hope you will enjoy.

Best of luck and good wishes!

Sincerely,

Neil A. Armstrong

Sincere thank-you from the mother

September 16, 1969

Dear Mr. Armstrong:

On August 28, 1969, you wrote a letter to my son, Douglas Ray
Conway, and enclosed an autographed picture of the crew. This is to
acknowledge your kindness with heartfelt gratitude. Douglas wrote you
that hearing from you would be the happiest moment of his life. Perhaps
it was. Just two hours after your letter arrived, he fell sweetly and peacefully asleep with the sweetest little smile on his face, not to awaken again
on this earth. A copy of your letter and the picture were buried with
him; the originals are framed beside a picture of Douglas in our home.

Everyone who knew and loved Douglas was so happy to see his
most important dream come true, and everyone felt like my oldest son,
who is a senior at William & Mary College, when he remarked on the
"wonder" of living in a world where men, as bound up in world affairs
as you must be, still take time out to attend to matters of the heart, a
child's heart.

My prayers and blessing will follow you, and men like you, always!
(And I wouldn't be a bit surprised if somewhere *you* have a new little
guardian angel who joins me in these sentiments.)

Sincerely yours,

Nancy M. Conway
Senior Clerk

"WE HAVE NO SUCH PHOTO ON FILE"

July 27, 1970

To Mr. Neil Armstrong—

Enclosed is a copy of an ad in our Christian Science Monitor which I answered with money and send to the A.T.A. Inc., San Francisco. I have had no reply to my inquiries from that firm.

I have also written the Space Administrations in Houston and Washington, D.C. to see where I can find a picture of Neil Armstrong taking the first steps on the moon. The suggested I write to James R. Dunlop Inc. which I did sending the enclosed copy. They wrote quote, "This photo is apparently a cropped portion of one of the other photos taken on the moon. We have no such photo on file."

Surely there is one available for the whole world saw Neil Armstrong taking that first step by television.

Please reply as to where I can find this print.

Sincerely,

Mrs. Helen Havens
Orlando, Florida

In Neil Armstrong's printed handwriting the following was written on the back of the original copy of Mrs. Havens's letter: "SINCE THE CAMERA WAS STILL IN THE LM COCKPIT, NO PICTURES WERE TAKEN EXCEPT TV AND 16 MM FROM THE RIGHT HAND WINDOW QUITE CLEAR BECAUSE OF THE POOR LIGHTING IN THE LM SHADOW." The original hand-writing was in a mix of uppercase and lowercase letters.

It was on the basis of this information provided by Neil that the letter below was prepared for reply to Mrs. Havens.

Reply from the chief of the Audio Visual Branch at NASA Headquarters

September 3, 1970

Dear Mrs. Havens:

In response to your letter of July 27 to Neil Armstrong, the photograph that appeared in the ad you submitted was taken off of a television monitor. I have enclosed two stills to show you the quality.

As you can imagine, there are literally thousands of pictures available from the television video tape. This exact picture could be obtained by screening the 16mm [illegible] that we have available at our motion picture depository. This exact frame can be obtained this way but it will be expensive. If you wish to pursue this, please let me know.

Sincerely yours,

Les Gaver
Chief, Audio Visual Branch
Public Information Division
NASA Headquarters
Washington, D.C.

"PLEASE DO NOT USE A RUBBER STAMP"

July 28, 1970

Captain Neil Armstrong, US Navy
National Aeronautics and Space Administration
Washington D.C.

Dear Captain Armstrong, Sir:

My apologies to you for intrusion on your time as I realize that you must receive many such requests as below.

If and when you have some time and its convenient to you to do so, could you please autograph your picture on the enclosed copies for me.

Like your self I work for USA in the Defense Supply Agency and I worked on the first instruments made back in 1961 and 1962 for Apollo, that is flight guidance systems at Lear Seigler and then I became Resident Rep. at Barden Bearing Corp. at Danbury.

I would very much appreciate your consideration. If you can, please make the autograph for or to me. I want to have them in my office.

Please do not use a rubber stamp.

THANK YOU SINCERELY,

Alexander F. Nahas
Danbury, Connecticut

"PLEASE INSERT CLAY FOOTPRINT INTO PLASTIC BAG"

August 7, 1970

Dear Sir:

Will you please make one small step for Jan?

My son, Jan, who is quite a fan of yours (myself included) would like to receive an impression of your left foot which you immortalized for enduring posterity, when you, as the first human alighted upon that cosmic temptress—the Moon, and thereby joined mankind in a celestial marriage whose offspring shall be those distant jewels of eternity—the heavenly bodies.

Thus far the achievement of the APOLLO MISSION has yet to be fully fathomed for it's inevitable proportions when man's journey resumes toward the many leagues beyond the present veil of the cosmos and our transitory grandeur.

Your participation with the APOLLO crew enraptured me to an akward attempt to translate my feelings poeticaly; and the enclosed poem is the result of that inspiration written during your voyage toward the Moon rendezvous on 20 July 1969.

I have provided within this package moist clay for your footprint, and if you please, using a scriber of sort, your autograph on the longitudinal axis of your footprint; also enclosed the APOLLO ELEVEN commemorative stamp, again, if you please, your autograph on the print side. Please insert clay footprint into plastic bag provided upon completion of doing your "Luna thing", taping extended opening to top side of package. The enclosed check is to defray costs for returning package in same container, if possible.

It is hoped that your recent appointment within N.A.S.A. will be challenging and rewarding, since your treasured knowledge will enchance future missions immensely.

Most Respectfully,

John Fedock
Shippensburg, Pennsylvania

"APOLLO ODYSSEY"

Strewn among radiant jewels of infinity,
Our celestial hope of yearning;
Since time began, still burning
In spacious flight. O vigil light.

Majestic are they in flight
From dawns creation to the brothers Wright;
From the Lone Eagle's Atlantic solo;
Now—the millenium, journey of Apollo.

Alas! Now, the temptress within grasp,
And anon, together with the past,
Forge the present, bringing the future near;
Closer, ever closer, his new frontier.

Where do we go into the new frontier;
Launched from embryonic hopes and fear,
From the seas of crises and desolation;
To the seas of tranquility and nobility?

John Fedock
20 July 1969

Reply from S. B. Weber, Neil's assistant

October 9, 1970

Dear Mr. Fedock:

Please accept our apologies for the delay in responding to your letter
to Mr. Armstrong. Because of his very heavy schedule of activities for
NASA, Mr. Armstrong is unable to respond personally to all of his mail
at this time. We regret that he is unable to respond to your request for
an impression of his "immortalized" foot. Requests of this nature have
become so numerous that it is not feasible to honor any of these. We are,
therefore, returning your check, along with the clay.

Since the stamp is too small to easily write on, we are sending you an
autographed picture of Mr. Armstrong under separate cover. Your stamp
is returned herewith.

Mr. Armstrong appreciated your poem "Apollo Odyssey" and has
asked us to convey his sincere thanks for your thoughtfulness in sharing
it with him.

Best wishes and, again, thank you for writing.

Sincerely,

S. B. Weber

"IF YOU EVER COME TO ROANOKE . . ."

September 8, 1970

Dear Neil,

I'd like to wish you a very happy birthday. I hope that you like the birthday card and that you can use the handkerchiefs. I sent along a picture of myself, so you can see what I look like. I recently graduated from high school in June. I have light brown hair and blue eyes. I live in Troutville, which is near Roanoke. You should visit Virginia. I think it's very beautiful here, especially in the spring and fall. Knowing what I know about you, I would say you would love it here. If you ever come to Roanoke, you have a standing invitation to stay at our house.

I enjoy reading about space missions and the astronauts. I keep a file on each mission as it comes up. My largest file, of course, was Apollo 11. I can remember the excitement I felt when Eagle touched down on the moon. I cheered when you made the first step on the moon. I read that you expect tourists to go to the moon. When they go, I'll be one of the first.

Congratulations on your appointment on July 1, 1970 as Deputy Associate Administrator for Aeronautics with NASA. I hope you saw the CBS News Special on July 21, 1970. It was a review of Apollo 11's triumphant mission. I hope there will be more specials such as this one.

Your first name: Neil, means a champion. Alden means friend. Your birth sign is Leo. You are very courageous according to your sign. You are a champion to me. I also like to think of myself as your friend. Even though I've never met you in person or spoken to you, I've watched you on television and read about you so much, that this is the only way I could possibly feel about you.

I realize how busy you are, but if you could find a picture of yourself that I could frame and send it to me, I would treasure it.

I can only thank you for being as you are: One of the bravest men of all time, and yet one of the sweetest, most considerate men alive. Please don't ever change.

Love always,

A Friend
Kathie Love
Troutville, Virginia

"HER PARENTS WILL NOT LET HER
NEAR ANY TYPE OF AIRPLANE"

September 30, 1970

Dear Mr. Armstrong,

I would like to ask your assistance on a very important matter.

Two weeks ago, on my 16th birthday, I got my private pilots liscence for glider's. I have been waiting to years for my liscence so I could take my girlfriend for a glider ride.

However, a complication has arisen. Her parents will not let her near any type of airplane. I have told them about the great strength and safety of sailplanes but they are not convinced.

I think that if someone as widely respected and trusted for good judgement as you are would write to them and explain, briefly, the virtues and safety of soaring, that it would do much to change their attitude.

If you agree to write to them I will send you a stamped envelope with their name and address on it.

Any help you could give me would be very greatly appreciated.

Yours hopefully,

Jim Shafer
Chalfont, Pennsylvania

"I HAVE BECOME QUITE CONCERNED ABOUT HIM"

October 23, 1970

NASA
Washington, D. C.

Dear Sirs:

Several weeks ago, after reading that Mr. Neil Armstrong had been made head of NASA, I dropped a postal card to him. I asked him to inform me just how much the ordinary citizen in this country will benefit from the space program to date and to justify the enormous cost. I also inquired about future plans and the justification for any further space flights.

I am a school teacher and since I have not heard from Mr. Armstrong after all this time I have become quite concerned about him. I can't help wondering if the reason that I haven't had any response from him is because he is not able to read or write. Perhaps there is some kind and literate secretary around who can read my letter to Mr. Armstrong and then send his reply to me. I really would like to have answers to my questions.

Sincerely yours,

Helen B. Lippincott
(Mrs. Howard Lippincott)
Riverside, California

Reply from S. B. Weber

November 14, 1970

Mrs. Howard Lippincott
Riverside, California

Dear Mrs. Lippincott:

Thank you for your letter of October 23 and your postal card of several weeks prior. NASA has attempted to demonstrate to the citizens of this country the benefits and potential benefits of the space program to date. Our budget and Congressional hearings document our accomplishments to date and the potential benefits that we envision from space flight.

In addition to these accomplishments, NASA is also responsible for conducting a research and advanced technology program to support air transportation. Mr. Armstrong, who is the Deputy Associate Administrator for Aeronautics, is taking the lead in providing the Nation with an aeronautics research and technology program that does benefit the ordinary citizen. Many of the benefits of this program have also been documented in Congressional presentations which are available from the U.S. Government Printing Office.

Additional inquiries relative to NASA's contributions may be directed to the NASA Public Affairs Office.

Thank you for your interest in writing and I hope you will be understanding of the length of time required to answer the deluge of mail received by Mr. Armstrong.

Sincerely,

S. B. Weber

"IN THIS SIMULATED FLIGHT . . . I AM MISSION COMMANDER"

October 20, 1970

Dear Mr. Armstrong:

I want to thank you for answering my letter of July 21 to you. I appreciate that you took the valuable time from your busy schedule to answer my letter.

In this letter, I have two things I would like to ask you. First, could you please describe your new position a little more clearly? I am greatly interested in the ground work that Nasa does, and I would be grateful if you could provide some extra information.

This December, I am going to simulate a lunar landing mission with two of my friends, Rob Tolley and John Sheppa. In this simulated flight, Rob Tolley is Lunar Module Pilot, I am Mission Commander, and John Sheppa will be Command Module Pilot. Rob and I are going to make a landing at Copernicus Crater and make three moonwalks, while John remains in orbit.

In preparation for the flight this winter, I am going to write out an extensive flight plan, and I was wondering if you could help us by giving us a few ideas and/or suggestions to help make the simulation more real? We have chosen "Cherokee" as code-name for the Command Module and "Apache" for the Lunar Module. We are scheduled to lift off on December 24, 1970 making our landing on December 29, with the return to Earth on January 3, 1971.

I would like to send a copy of our flight plan to you as soon as it is finished, so you could look it over and then return it with more suggestions or corrections. With this letter, I am also enclosing a copy of our crew's flight patch.

I remain,

Yours, respectfully,

David Shaw
London, Ohio

Dear ArmStrong
could you Send me a
picture of the moon
and a model of
a spacecapsule

From
Rusty Buckmaster
124 W. W th St
Port Angeles, Wash 98362

No Pic
(s) Put
sent 4-3-7~

3-15-7~

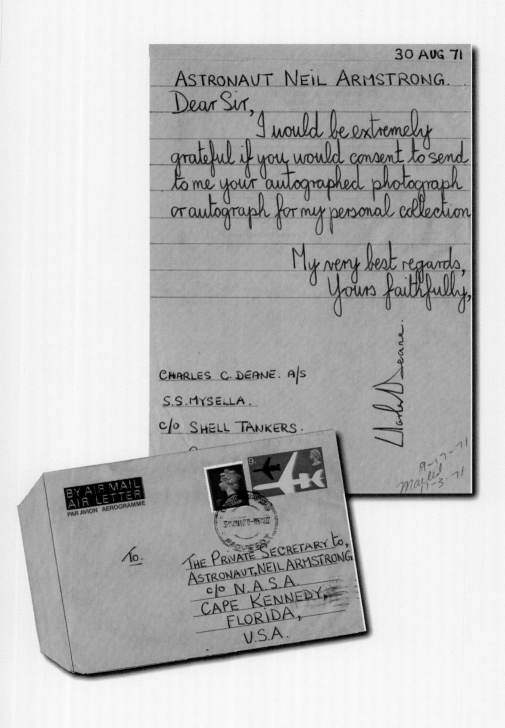

30 AUG 71

ASTRONAUT NEIL ARMSTRONG.

Dear Sir,

I would be extremely grateful if you would consent to send to me your autographed photograph or autograph for my personal collection

My very best regards,
Yours faithfully,

CHARLES C. DEANE. A/S

S.S. MYSELLA.

c/o SHELL TANKERS.

BY AIR MAIL
AIR LETTER
PAR AVION AEROGRAMME

To. THE PRIVATE SECRETARY to,
ASTRONAUT, NEIL ARMSTRONG
c/o N.A.S.A.
CAPE KENNEDY,
FLORIDA,
U.S.A.

ARMSTRONG AND ALDRIN ON MOON

APOLLO 11

by William Evans

Dear Mr. Armstrong,

    I think you are a very nice man. After my bir-
thday i found out that you where the first man on
the moon. When is your birthday? Mine is December 17.
Do you think it will be possible for you to write to
me? Would you please send me your picture? I would real
-ly appreciat that very much. Me and my friend thout
you were in your 40s. Is that true? If your wondering
how old I am , Im only 8.

                yours truly

                Christine Granger

(handwritten) Neil

(handwritten) postmarked Nov. 14, 1972

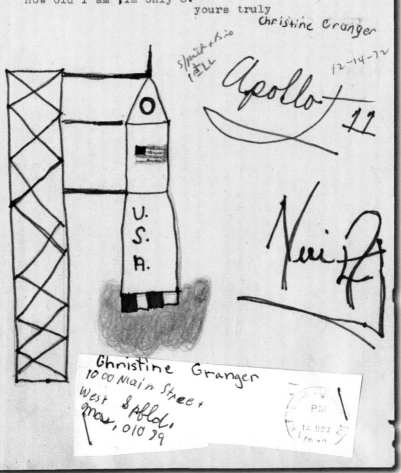

(handwritten annotations on drawing) s/print + bio 1ZL · Apollo 11 · 12-14-72 · Neil *(signature)*

Christine Granger
10 00 Main Street
West Spfld.
Mass, 010 89

CLIFFORD HARMON VALLONE

9 Wilde Avenue, Apt. 5
Drexel Hill, Pa. 19026
24   July   1 9 7 1

Mr. Neil Armstrong - Code RD-A
National Aeronautics and
Space Administration
Washington, D. C. 20546

My dear Mr. Armstrong:

At Christmas 1970, I wanted to send you my Christmas
Card but I did not know where to send it.

I just learned of your address, and, while I am
tardy, I take pleasure in enclosing my 1970 Christmas
Card which I have created.

The picture in the background is that of Clifford Harmon,
pioneer aviator and balloonist. He also established the
Harmon International Trophies. I am proud to say that
I am his namesake. He and my father became friends early
in 1900. I have carried that name for many years with
a great deal of pride, dignity and respect.

I know that you received the Harmon Trophy, and I am
proud of you as an incorruptible patriot. Your performance has been excellent. Men of your caliber are
scarce. We need more like you!

May health, happiness, and God's protection accompany
you everywhere you go.

Sincerely,

Clifford Harmon Vallone

*Season's
Greetings
and Best Wishes for a
Happy New Year*

*

Clifford Harmon Vallone

western union   **Telegram**

NYD253(1934)(2-030341C292002)PD 10/13/76 1926

TWX WHITEHOUSE WSH

019 DLY GOVT WHITE HOUSE DC OCT 18

PMS MR. F. D. HALL, PRESIDENT, DLR DONT DWR

THE WINGS CLUB

THE AMERICANA HOTEL (C/O ASST HOTEL MNGR ON DUTY)

NEW YORK, NY 10019

I EXTEND WARMEST GREETINGS TO THE MEMBERS OF THE WINGS CLUB
AND TO YOUR HONORED GUESTS AT THIS THIRTY-FOURTH ANNUAL DINNER.

I AM DELIGHTED TO SHARE IN YOUR SENTIMENTS AS YOU PRESENT
THE CLUB'S DISTINGUISHED ACHIEVEMENT AWARD TO NEIL ARMSTRONG.
AS THE FIRST OF TWELVE AMERICANS TO WALK AND RIDE ON THE MOON,
NEIL ARMSTRONG MADE AN ENDURING CONTRIBUTION TO THE ADVANCEMENT
OF SCIENCE AND TO THE NOBLEST SPIRIT OF HUMANITY. IT IS MOST FITTING

SF-1201 (R5-69)

---

western union   **Telegram**

THAT THIS AUSPICIOUS OCCASION HAS BROUGHT TOGETHER THE KEY
MEMBERS OF THE TEAM THAT SUCCESSFULLY EXECUTED THE GREAT
MISSION WHICH WON HIM WORLD-WIDE ACCLAIM.

I HOPE THAT IT WILL BE A MOST MEMORABLE AND HAPPY OCCASION
FOR NEIL AND FOR THE MANY DISTINGUISHED COLLEAGUES AND FRIENDS WHO
ATTEND. HE HAS EARNED THEIR RESPECT, THE GRATITUDE OF OUR
NATION AND THE ADMIRATION OF PEOPLE EVERYWHERE.

GERALD R. FORD

NNNN

SF-1201 (R5-69)

BARRY M. GOLDWATER, JR.
27TH DISTRICT OF CALIFORNIA

SCIENCE AND ASTRONAUTICS
COMMITTEE

SUBCOMMITTEES:
ADVANCED RESEARCH AND
TECHNOLOGY
NATIONAL BUREAU OF STANDARDS

Congress of the United States
House of Representatives
Washington, D.C. 20515

WASHINGTON OFFICE:
LONGWORTH HOUSE OFFICE BUILDING
(202) 225-4461

DISTRICT OFFICE:
10850 RIVERSIDE DRIVE
SUITE 404
NORTH HOLLYWOOD, CALIFORNIA 91602
(213) 769-0727

January 30, 1970

Mr. Neil Armstrong
Astronaut Affairs Office
Code CB
NASA Manned Spacecraft Center
Houston, Texas 77058

Dear Neil:

Well, we are off and going on another year of excitement here in
Washington. After sitting in on a panel on Science and Technology,
I couldn't help but recall my very exciting experience at Cape
Kennedy watching Apollo 12 leave the earth; and, of course, this
excitement was projected even higher when you took the time to
describe what was happening. You were very kind, Neil, to offer
your explanation and I can't thank you enough, for it made the
whole event so much more interesting and informative.

I have seen you from time to time on television or in the news
media performing for one cause or another. The highlight, I am
sure, was going to Viet Nam with Bob Hope. I just want you to
know that I think that you are doing a fantastic job in your
service to the country, and I would like to join the rest of the
nation in offering my congratulations and thanks. The type of
spirit and leadership that you represent is the kind of thing
that is needed to challenge Americans into appreciating this great
country of ours.

I hope to see you sometime soon in the future. If you are ever in
Washington or out in "beautiful downtown Burbank", please be sure
and say hello. Again, thank you for all your kindness, and I certainly wish you the best of fortune in the year 1970.

Sincerely yours,

BARRY M. GOLDWATER, JR.
Member of Congress

30.9.72

Dear Neil Armstrong.

I am a girl from Israel, and
I have a big collection of
autographs. I'll be very
happy if you'd be so kind
and send me your autograph.

   I wish you the best,

     Osnat Yefe-Nof.

     אסנת יפה-נוף.

My address is:
Osnat Yefe-Nof
Bet-Berl, Kfar-Sava
44 925
ISRAEL

no Pi.
5/pct
SENT 2/22/KP

to:
Mr. Neil Armstrong (The first
                man on the
                moon.)
N.A.S.A

HOUSTON-TEXAS
U.S.A.

16-10-72

Mr. Armstrong,

You did the great work and with this work you opened the gate of sky on the men. We all thank you for this great work which you did that.

Miss Hoora-z iaeepour   14 years old.

1971-7-11   7-26-71

DEAR MISTER *Armstrong,*

I'M ITALIAN BOY AND I FOLLOW WITH
INTEREST THE SPACE ADVENTURE.
IN MY YOU ARE AN

AND I WOULD LIKE VERY MUCH
RECIVE AN YOUR AUTOGRAPH ON
THE BEHIND OF AN YOUR VISITING
CARD MY ADRESS IS :

BOB NOTO
C.SO BRAMANTE 65
10126 TORINO
ITALY

Thank you very much !

Auto
Photo
Sent 12/1/70

(BOB
★
NOTO)

P.S. ESCUSE MY
MISTAKES!

Arni,
10th July, 1970.

From:
M. Gunasekaran,
State Bank of India,
Arni. (N.A.DT.)
Tamil Nadu, South India,
India.

To
Neil A. Armstrong,
C/O. The Chief of Public Affairs,
John F. Kennedy Space Centre Nasa,
Kennedy Space Centre,
Florida 32899 U.S.A..

Dear Armstrong,

          With all my heart, I love you and respect you.
When you were in Apollo 11, I felt as if I was with you
all the way to the moon. Every day, as soon as I get
up from bed, I look at your photo mailed on the wall
in front of me and gather strength and courage for my
day's task. Inside my shirt pocket, I have another
photo of yours, which I gaze at when alone or away
from home.

          Oh how lovely you are Mr. Neil, the bravest
and the most courageous of men! I hail this day 20-7-70,
as a mark of your setting foot on the moon a year ago!

          How eagerly I wish for your autograph!

                              Your ever remembering friend,

                              Karan
                              10-7-70
                              (M. Gunasekaran)

Auto. Photo
Sent 10/30/70

BY AIR MAIL
PAR AVION
हवाई पत्र
AEROGRAMME

RD-A

Mr. A. Armstrong,
C/O. The Chief of Public Affairs,
John F. Kennedy Space Centre Nasa,
Kennedy Space Centre,
Florida 32899 U.S.A.
U.S.A.

SECOND FOLD

Santa Catalina School
MONTEREY · CALIFORNIA

February 8th, 1971

Commander Neil Armstrong
National Aeronautics and
    Space Administration
Washington D.C. 20546

Dear Commander Armstrong,

    I am very interested in persuing a career
in space science and the space program. I would
be very grateful if you would be kind enough to
take the time to consider what types of work I
should persue during my summer vacations in
order to arrive at the maximum training and
experience for space work by graduation date
from university.

    I am completing my junior year of high
school and hope to work in the Lunar Receiving
Laboratory at Houston next summer. This is in
the process of being arranged during the next
two months. Do you have any suggestions as to
any N.A.S.A. position that would give me further
experience during the summer of 1971 or a future
summer?

    My interest in our space program and in
space science has lead to many great experiences
for me. I have toured several factories in the
Los Angeles area which make Apollo spacecraft
parts, visited the Jet Propulsion Laboratory at

Cal Tech., saw two Apollo launches, worked in
the geology labs of the University of California
at Los Angeles, and went to the Houston and
Huntsville Space Centers with a motion production
crew. The more I learn about our space program,
the more I want to become a part of it. I
enclosed a summary of some of the things that I
have done because of my deep interest in space.

I would very much like to have an appointment to come and confer with you on Monday April
12th, while I am on a college tour in the East,
or during the month of August 1971 whenever you
would be in your office.

Do you know of any present or future program
for female astronauts? Will there be women on the
fifty-man space station? I hope to someday
become an astronaut if possible. Otherwise I
would participate in our space program in some
field of science research.

I would be interested in corresponding with
anyone that you would suggest. I would be most
grateful to get a letter from you with both your
suggestions and a possible appointment date in
the Spring or Summer.

Sincerely yours,

Dana Turner

Miss Dana Turner
Santa Catalina School
Mark Thomas Road
Monterey, Calif. 93940

Dear Mr. Armstrong I wish
I wish you would come to
The school and show The
equipment so everyone can see.
I hope you could bring some
equipment. please will you come?
Margo Mechen

School in Rose Valley
Moylan, Pa.

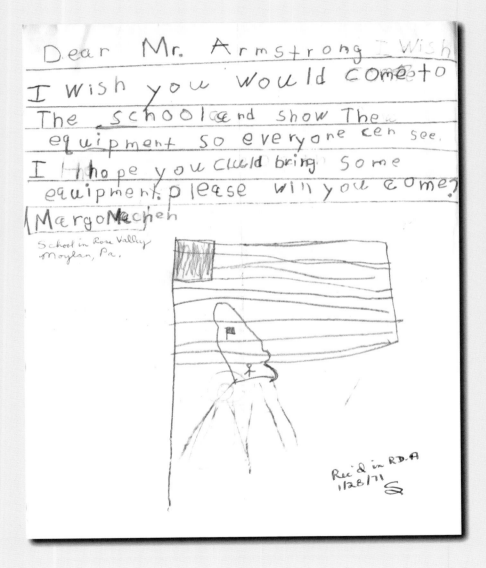

RD-A

Margo Machen
School in Rose Valley
Moylan, Pennsylvania

Dear Margo:

Thank you for your very kind letter to Mr. Neil Armstrong.
He is always pleased to hear from young people with a
keen interest in our space program. Unfortunately,
because of the many letters he receives, he is unable
to personally reply to you at this time.

Mr. Armstrong appreciates your kind invitation to visit
your school; however, because of his heavy schedule of
activities with NASA, he regrets that he will be unable
to accept.

We are enclosing an autographed picture of Mr. Armstrong
which we hope you will enjoy. Best wishes and, again,
thank you for your interest in writing.

                        Sincerely,

                        /S/

                        S. B. Weber

Enclosure

SWeber 1/28/71

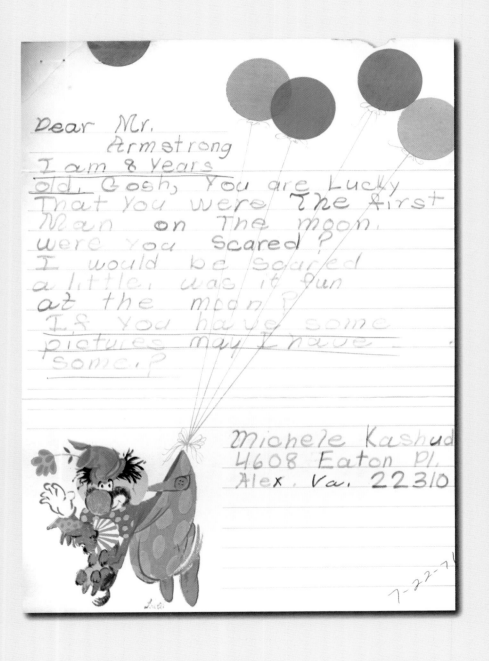

Dear Mr.
        Armstrong
I am 8 years
old. Gosh, You are Lucky
That you were The First
Man on The moon.
were you Scared?
I would be scared
a little. was it fun
at the moon?
If you have some
pictures may I have
some.?

Michele Kashud
4608 Eaton Pl.
Alex. Va. 22310

7-22-7

Enclosed with the letter was a drawing of the flight patch with the words "APOLLO XVII—SHAW—SHEPPA—TOLLEY" and "Cherokee" and "Apache."

"MY AMBITION IS TO BE AN ASTRONAUT, THE FIRST GIRL ON THE MOON"

December 26, 1970

Dear Mr. Armstrong,

I would like to thank you for the photographs and literature I recieved from NASA. I enjoyed them very much. Everyone I showed the pictures to, liked the one taken from Apollo 8, showing the moon in the foreground and the Earth in the background, best. My favorite is almost like that. The only differance is the LM coming up from the moon's surface. I didn't get this picture in the package.

My ambition is to be an astronaut, the first girl on the moon. I doubt if I'll make first girl on the moon, because someone will probably beat me to it. I am 12 years old and make straight As in school. I first wanted to be an astronaut when Apollo 11 went up, and wanted to be one ever since. I'm making a scrapbook about space, it's two volumes long, so far. By the time it's finished it will probably have about four. There is one thing stopping me from becoming an astronaut, I wear glasses. The eye doctor said that I will have to wear them all my life, but my teacher keeps telling me I probably won't, so it is hard to say.

You are my hero. I like all the astronauts but you're the best. You did thing that is now my goal. You were the first person on the moon. I would have given almost anything to have gone up on Apollo 11. A lot of people feel that the lunar module is ugly. Personally I think that it is cute. I'm not saying it's pretty, but it is cute.

I'm going to Purdue to college, and major in astro-physics, minor in geology. I'm going to get my masters at Purdue and my doctorate at MIT. My teacher, Mr. Shafer, gives me several things on space and about physics. He says that I'm always asking him questions he can't answer, like what is an MEV. I have my brother's physics book from college. It has a lot of information in it. My brother is planning to teach physics. I have another brother Paul living in Conn. My youngest brother, Larry, goes to Ball State. He plans to take me to the planetarium at Ball State sometime. I'm also planning to go to the observatory at Carlam College some time. I got a moon globe and a star finder for Christmas, and using my dad's telescope to look at the moon.

You don't have to answer this letter, but it would be appreciated if you would.

Yours truly,

Debbie Schuler
Milton, Indiana

Reply from S. B. Weber

January 6, 1971

Dear Debbie:

Thank you for your recent letter to Mr. Armstrong. He is always pleased to hear from young people with a keen interest in our space program. Unfortunately, because of his heavy schedule of activities with NASA and the tremendous volume of mail he receives, he is unable to personally respond to your letter at this time.

We are glad to hear that you received the various items mailed to you and seemed to derive so much pleasure from them. We trust that your interest in our activities will continue as our activities progress.
With best wishes,

Sincerely,

S. B. Weber

"MY PARENTS HAD TO POUR COLD WATER
ON MY FACE TO KEEP ME AWAKE"

January 10, 1971

Dear Mr. Armstrong,

I am 10 years old and in the 5th grade. My parents are Missionarys here
in Okinawa.

   We are studying space science, and I wondered if you could send me
[illegible] on space.

   Last week we had a moon rock here on display at Stillwell Fieldhouse.

   We were living in the States when you made the Moon Flight. My
parents had to pour cold water on my face to keep me awake but I saw
it! And I'll never forget it!

Signed,

Marianne Madden
"On Okinawa"

"TOWARDS AN ECOLOGICAL AWARENESS"

February 6, 1971

Dear Mr. Armstrong,

Let me belatedly congratulate you and the NASA team for your magnificent accomplishment. It appears as a singular beacon, which hopefully
will cut through the darkness of self-serving national interests and
encourage mankind to unite into a global consciousness.

   Today the United States leads the world's attention towards an ecological awareness, which necessarily must reorder priorities, and restate
man's position in Nature's grand order. Today we are riding a crest of

social consciousness towards environmental degradation, and yet, as hundreds of millions of people in Earth's fragile biosphere follow the course of Apollo 14's return home, they are also witness to what appears to be a senseless, and degrading act of our all too typical over consumptive society of disposable goods. I refer to the lunar crash-landing of the L.E.M., spewing its debris and dubious bacteria across an otherwise virgin landscape. Hypocracy is not new to us, and yet what excuse can we offer to the people who have paid for the mission, and especially to those whom we inadvertently impress? Would it not be more consistent with President Nixon's expressed concern for the environment to direct the L.E.M. into the Sun, where we could rest assured the material would be recycled? Then again it seems a terrible waste not to salvage the L.E.M., though I can understand how this might not be economic. Would it not be economic to at least rescue the first stage rocket for reuse?

Of course I would appreciate your comments on the subject and your endeavor to rectify the situation. And if you like, I'll send copies of our letters to the major newspaper in Philadelphia and the University of Pennsylvania newspaper.

Sincerely yours,

R. Tek Nickerson
Grad. Student of Ecological Regional Planning
Philadelphia, Pennsylvania

Personal reply from Neil

April 13, 1971

Dear Mr. Nickerson:

Thank you for your letter of February 6, and please accept my apologies for the delay in responding.

NASA has taken all reasonable precautions to avoid disrupting the natural environment of the moon. Proposed experiments on the lunar surface are reviewed by scientific specialists and advice is solicited from the best scientific personnel available.

For example, the experiments involving impact on the moon by the spent Apollo IX Ascent Stage, to which you refer, was endorsed by the Space Science Board of the National Academy of Sciences. The purpose is to produce seismic events of known location and intensity which can be detected by seismometers deployed during the same and earlier Apollo missions. The resulting seismic waves probe some tens of kilometers beneath the layering in the vicinity of the seismometer. It is noted that the energy of these impacts is infinitesimal when compared with impacts of extremely large bodies that have struck the moon already. The energy of Apollo components is so low that we cannot observe the small craters formed by them. Compare this with large impact craters like Tycho and others many miles in size on the moon.

You may recall that we did send the LM of Apollo 10 toward the sun when we had finished with it. At that time we did not have any seismometers on the surface so impacting the moon would have had no value.

Your suggestion that the Lunar Module and first stage booster be reused has been considered. In a continuing effort to cut costs and economize where possible, NASA has always considered the reusability of expensive hardware as a desirable objective. Early in the Apollo Program various modes of operation were considered and evaluated. While reusability of major hardware components was considered, this would have further complicated an already tremendously complex system and thereby reduced our chances of success. Therefore, our first generation of spacecraft and boosters are designed primarily for reliability, not for reuse. Additionally, we have examined the cost of modifying existing first generation space hardware for reuse and find that the cost of modifications would far exceed the potential cost savings.

The reusability of space hardware is being given very serious attention in the planning of our future program. The Space Shuttle is expected to reduce by about a factor of ten the cost of launching objects into earth orbit. Thus, with this possible launch vehicle, the cost of boosting payloads should drop from an average figure of $1000.00 per pound to about $100.00 per pound.

Hardware costs must be reduced if we are to proceed in space at a meaningful pace and NASA is looking carefully at all possible means of achieving this objective.

Sincerely,

Neil A. Armstrong
Deputy Associate Administrator for Aeronautics
Office of Advanced Research and Technology
NASA Headquarters
Washington, D.C.

"HURRY DOWN SOON"

February 10, 1971

DEAR MR. ARMSTRONG,

I was watching The Courtship of Eddie's Father tonight and it was about
when Eddie had Gordon Cooper over for lunch. I want to know if you
can come down to Paducah and have supper with me. If you can come
to [gives his home address].

Now I will tell you how to get to my house, catch a cab and tell him
that you want to come to [gives street address]. YOU will have to tell
him to turn at albrittons drug store go down West Jefferson then turn
left and go down Pines Road. Then tell him that he is supposed to turn
left and it is the last house on the left.

HURRY DOWN SOON.

SINCERELY,

Bobby Martin
c/o Paul Martin
Paducah, Kentucky

"COME TO MY HOUSE TO HAVE DINNER WITH ME"

February 16, 1971

Dear Mr. Armstrong,

Hello! How are you? I would like you to come to my house to have dinner with me, Please? I am 8 years old. I live at: [gives address and telephone number].

Love,

Pamela Grapatin
Geneva, Ohio

P.S. My house is *yellow* and it is by a yellow sign like this: [drawing of a diamond shape with "School Crossing" written inside]. My House is on rt. 84 near the long hill or space on the other side of the road.

"I LIKED WHEN YOU PUT YOUR FOOT ON THE MOON"

February 22, 1971

Dear Neal Armstrong,

My name is Thomas Michell. I live at [gives address]. I would like to know if you would have time to have lunch with me along with no camers so that you are not on T.V. I liked when you put your foot on the moon. I am sorry that this is not on good paper.

Your Friend,

Thomas Michell
Staten Island, New York

"I TOLD HER THAT SHE WAS WRONG"

March 3, 1971

Dear Mr. Armstrong

I'm writing to you because I have a problem. My Science teacher Mrs. Billings was showing us a old movie on space exploration. After the movie was over, we were talking about the first trip to the moon. She told us that she did not belive that you ever set foot on the moon, and that the whole thing was faked. I told her that she was wrong because they wouldent spend so much money on something that was fake. She said that I or anyone else would have to give her some proof before she would believe me. So I'm writing you to ask you to send me a letter or something that I can show her so she will believe that you did land on the moon.

Signed

J.C. Bussard
Kenosha, Wisconsin

Reply from S. B. Weber

March 8, 1971

Dear J. C.:

Thank you for your letter to Mr. Neil Armstrong. He is always pleased to hear from young space enthusiasts. Unfortunately, because of his heavy schedule of activities with NASA, he is unable to personally reply to each of the many letters he receives.

We wish that we could assist you in convincing your teacher that our astronauts really did walk on the lunar surface. However, if she does not believe after having witnessed it through the live telecast, a letter from Mr. Armstrong certainly would not validate the action.

We do appreciate your interest and hope that it will continue as our activities progress.

Sincerely,

S. B. Weber

"YOUR INTERPRETATION OF LIFE WOULD BE MOST INTERESTING"

March 3, 1971

Dear Mr. Armstrong:

I am conducting an English class project in which I am to ask various people of various professions to explain "what the word *life* means to them." Since you were the first man to step foot on the moon, I feel that your interpretation of life would be most interesting. I know that your schedule is extremely full, but your reply to my letter by at least April 18 would be most appreciated.

Thank you,

Cheryl D. Pillson
Age 17
New Castle, Indiana

Note to Neil from Geneva Barnes

April 27, 1971

Mr. Armstrong:

In a case like attached, do you wish to send an autographed
photo, or not?

    The girl is 17, doing an English class project on what the world life
means to various people.

    I wrote her the "cannot honor your request because of the demands
on his time" letter (attached).

Gennie

*At the bottom of Geneva Barnes's note to Neil is Neil's reply to her in his printed
handwriting: "No—I guess not. Thx."*

Reply from Geneva Barnes

May 5, 1971

Dear Miss Pillson:

Mr. Armstrong asked me to thank you for your letter. He also sends his
apology for the late response but your letter did not reach this office
from Houston until last week.

    He is always pleased to hear from young people and would like to be
able to respond personally to all and be able to take the time to answer
all of the questions he receives. His definition of the word life deserves
considerable thought to be responded to properly and, regretfully, Mr.
Armstrong is unable to honor your request for an answer because of the
demands on his time.

    Thank you for writing.

Sincerely,

(Mrs.) Geneva Barnes
Secretary to Mr. Armstrong

"AND MAY I HAVE THE FIRST AID KIT"

March 19, 1971

Dear Mr. Armstrong

I saw your trip to the moon in 1969. I knew it would be a success. I'd like to know if I could have your autograph and photograph. I'd like to have it for my colection of astronauts. So can I have your autograph and photograph.

Thank You

Eric Moss
Westbury N.Y.

And may I have the first aid kit you used on your trip.

"YOU CAN SPEAK TO AMERICA, AND IT WILL LISTEN"

March 30, 1971

Dear Col. Armstrong

In the past few years it seems that whenever money is needed for a government project, the first to receive budget cuts is the Space Program. While the Dept. of Defense spends billions, the Space Program's millions dwindle. Our Space Program has taught us more in ten years than any other program has in a hundred, but the Space Program is slowly dying. There is not much I, as a private citizen can do about this crisis, but you, and the other highly respected men of the Space Program can do something. You can speak to America, and it will listen. We must save this vital program, because if we stifle our scientific growth and curiosity of the unknown, progress will die, and we will die.

Sincerely yours,

Scott Siebels
Shawnee Mission, Kansas

Reply from Geneva Barnes, Neil's secretary

August 4, 1971

Dear Mr. Siebels:

Mr. Armstrong has asked me to thank you for your recent letter express-ing concern about the budget cuts in the space program, and urging him to speak to the public more often about this vital program.

Mr. Armstrong continues to make speeches stressing the importance of the nation's space program and the following are excerpts from three of those speeches:

In his remarks at the United States Military Academy on May 4, when he accepted the Sylvanus Thayer Award, Mr. Armstrong stated:

". . . those faltering steps of this first decade of space flight were a spectacle that has been unparalleled in history. Our knowledge of the planet has increased a hundredfold; our understanding of the moon—a thousandfold and more. However, each new discovery has uncovered a new question. Progress, then it seems, is answering the questions of those who have gone before and providing new questions for those who follow.

"The first decade of space flight was a time of exploration. The next decade will likely be a time of application. Emphasis will be placed on how to reap the benefits from our new-found knowledge. This is consistent with the general contemporary tendency in our nation to turn inward and concentrate on human needs and creature comforts. Our space investments are being called on to emphasize the applications of this new technology to serve our people. And this is very appropriate. I do hope, however, that we will continue to explore our universe."

From his Commencement Address at Ohio State University on June 11, 1971:

". . . Today, the United States cannot afford to neglect, as it did the airplane in the early days of aeronautics, a technology so powerful in its potential as manned space flight. Any technology that can take man to the moon and back less than ten years after the first pioneering orbit of earth is not a toy to be lightly cast aside when our attention is distracted by other matters. Manned space flight, the art of mastering the prime forces of the universe, in an arena vast beyond imagination, ought to be judged for what it is: an enormously powerful and versatile new capability of immense consequence to all mankind—too great for us to ignore or downgrade its significance for the future. A nation's, as well as an individual's, true wealth is measured by the capabilities, not the material possessions it may accumulate. Spacefaring, just as seafaring and aeronautics did earlier in history, is moving forward on the stage of human affairs, adding a new dimension to our economic, cultural and spiritual potential."

From his University of Maryland (Heidelberg, Germany) Commencement Address:

". . . Of course, critics have complained that we are doing too much in space, with so many unsolved problems here on earth. But the positive approach is not to do less in space, but to do more on earth—and do it better. We must continue our exploration and search for understanding of the universe around us while at the same time applying the lessons we have learned from our space achievements to other world needs. They certainly increase rather than decrease our hope, our ability, and our resolve to face and overcome new and chronic problems here on the surface of the earth."

We hope your interest and enthusiastic support will continue.

Sincerely,

(Mrs.) Geneva Barnes
Secretary to Mr. Armstrong

"DID YOU MEET ANY LITTLE GREEN MEN?"

July 1, 1971

Dear Mr. Armstrong,

Would you please send me some information about NASA? Could I also have an autographed picture of you?
    Did you meet any little green men?

Pam & David Nickerson
Ocean City, New Jersey

Reply from Geneva Barnes

August 18, 1971

Dear Pam and David:

Mr. Armstrong asked me to thank you for your letter. He is always pleased to hear from young people and to learn of their interest in the space program. He would like very much to reply personally to all, but

I'm sure you understand that this is not possible because of the demands on his time.

We are enclosing an autographed picture of Mr. Armstrong and some information which we hope you will enjoy.

Sincerely,

(Mrs.) Geneva Barnes
Secretary to Mr. Armstrong

P.S. Mr. Armstrong and Colonel Aldrin did not meet any little green men.

"HOPE TO SOMEDAY JOIN THE RANKS OF OUR SPACE PIONEERS"

July 3, 1971

Dear Mr. Armstrong,

I have long followed the Space Programs of the United States and the Soviet Union and hope to someday join the ranks of our space pioneers in the new adventure. After Apollo 13 and now Soyuz 11 the world begins to wonder about manned exploration of the new frontier, and bring up the inevitable question, "Is this trip necessary?" I have written many articles that have appeared in many newspapers concerning this very question and have even written a few short stories which I hope to have published. Yes, Apollo 13, Soyuz 11 like all missions, is necessary. Each mission increases man's knowledge of the Earth and the universe, and understanding these environments will help us to understand the future course not only of the Earth, but of its people. The brave Astronauts of Apollo 13 like all American Astronauts knew the risks of their profession and are also well aware of the unlimited potential of space exploration. I have high hopes of entering the Air Force Academy next year and begin my efforts of applying for the Astronaut Corps.

I followed your Gemini 8 and Apollo 11 flights with keen interest and I was extremely proud to watch you and Mr. Aldrin roaming the lunar surface. With your step on the lunar surface began a new leap forward for the human civilization, man is embarking upon an era of space expeditions.

Enclosed is one of my favorite picture of you taken I believe in the LM just prior to lift-off and I was hoping you would autograph it with your first words on the lunar surface: "That's one small step for a man, one giant leap for mankind," I would really appreciate it very much. Also enclosed is a copy of HOE (Help Our Earth) a school publication of which I was one of the editors.

Thank you

Sincerely,

Michael David Rose
Walnut Creek, California

Reply from Geneva Barnes

July 6, 1971

Dear Mr. Rose:

Mr. Armstrong has asked me to thank you for your letter and for your school publication. He is always pleased to hear from those who are interested in the space program and would like to reply personally to all. However, I'm sure you understand that this is not possible because of the demands on his time.

He was happy to sign your picture but did not inscribe his famous quote. Mr. Armstrong never quotes himself.

Sincerely,

(Mrs.) Geneva Barnes
Secretary to Mr. Armstrong

"I NAMED HIM BRIAN (NEIL) SMITH"

July 12, 1971

Dear Mr. Armstrong

My son will be two this month. He was born in Cape Canaveral Hospital when you landed on the moon. I named him Brian (Neil) Smith, after you.

   I thought it might be something he would cherish getting a card from you. I myself would consider it a great honor, as I work at the Cape.

Thank You

Jem Smith
Rockledge, Florida

"MOON DAY U.S.A."

July 12, 1971

Dear Mr. Armstrong,

I have intended writing this letter for sometime but never seemed to get at it. I shall be short and to the point so as not to take up your valuable time.

   In the summer of 1969 I wrote you and also sent you a recording of my song Moon Day U.S.A. I asked you in the letter if you would kindly listen to it and let me know what you thought of it. This I would appreciate very much. I know you were on tour at that time and I did receive acknowledgement of the lyrics (which I had sent to NASA and the record addressed to you) but nothing was mentioned about the record. I am wondering if you ever received it as I cannot think you would ignore my request for your evaulation of it thus I am prompted to write this letter. Let me say here I was amazed and shocked that the

Columbia Recording Company did not find enough commercial value in it to warrent their recording it. I only sent it to two record companies and the other company wrote a long letter complimenting me on it and requesting information as to the recording artist. As far as I know it is the only song recorded of the moon-landing. To me a record that was recording the greatest achievement in our history to be found worthless is incomprehensible to me. Also, I wish to state that I sent out two demonstration record and they came to nought—one was lost and the other was never acknowledged. The one not acknowledged was sent to the people of your home town as a gift and believe it or not I heard nothing—was sent in care of the newspaper of your home town. I retract, I did send one other record to another party because I felt it would go nice with the gift he had at that time received—a very well known personality and again I got acknowledgment of the lyrics but nothing was said about the record. I feel the party never got it as I had written him also explaining why I wanted him to have it. Well, that is the story about the song Moon Day U.S.A. Great numbers here in this university town for whom I played it thought it was great and one even wanted me to take it to Johnny Cash who was preforming here at that time—the recording artist sounded something like him.

Thank you Mr. Armstrong, for listening to me and I do hope you will favor me with a reply.

Did not know so near end of page so please excuse.

Sincerely yours,

Ann Houchin
Urbana, Illinois

May 1, 1972

Dear Mr. Armstrong,
I have written you several letters and up to date I have never received a reply. I shall try again. I wrote you at Washington D.C. and had a reply from the Public Affairs office saying they had forwarded my letter to you. All I wish to know, Mr. Armstrong, is whether you ever received

the record of "Moon Day U.S.A." I sent you just after the moon landing. I asked you to let me know how you liked it. I can't think that you
would just ignore me and I am trusting you will favor me with a reply
this time. The only answer I ever got was a reply from NASA saying they
had received the poem—poem not the record. I sent another record to
another prominent person and it was lost in the mail. Many Professors
here heard it and loved it but the young folks do not approve of spending so much money from the space program and as it is the young that
buy the records I had no success with a record company. However, I did
not try very maney.

Trusting that I shall hear from you soon, I remain

Sincerely,

Ann Houchin
Urbana, Illinois

Reply from Fern Lee Pickens, Neil's assistant

July 27, 1972

Dear Miss Houchin:

Thank you for your most recent letter to Mr. Armstrong. He is unable
to reply personally to the many letters he receives and he has asked me
to thank you for the lyrics, which you say were acknowledged by NASA;
but, I am sorry to tell you that the record of which you speak has never
been received by Mr. Armstrong.

An extensive search has been made of NASA files for your previous
correspondence to Mr. Armstrong, and for the record, unfortunately,
we have been unable to locate any information. I am aware, however,
that when Mr. Armstrong moved from Houston to Washington, D.C.,
some of his files and papers were misplaced. I can only conjecture that
your previous correspondence must be in the missing files. What could
have happened to the record is also conjecture since you say that NASA
never did acknowledge receiving it—only the lyrics. I am sure you can

appreciate that NASA and Mr. Armstrong were swamped with mail and gifts of all kinds from all over the world after the lunar landing. Although Mr. Armstrong could not and would not have commented on the relative merits of a particular composition, had he received the record, he, or a NASA staff member, would have surely acknowledged receiving it. We can only hope that someday, somehow, the Moon Day USA record will turn up, along with the missing files and papers.

We regret very much any inconvenience or discomfiture you have been caused by this unhappy circumstance. We appreciate your interest in the space program and hope your enthusiasm will continue as our nation's space efforts progress.

With kindest regards,

Sincerely,

(Mrs.) Fern Lee Pickens
Assistant to Mr. Armstrong

"HOW MUCH DOES IT COST FOR A LITTLE PIECE OF MOONDUST?"

July 26, 1971

Dear Astronaut Armstrong:

Could you please send me some pictures of the moon and outer space? I hope they are in color. I love color! The prints the newspaper turned out weren't the greatest things in the world. They didn't do justice. By the time you get this letter, more astronauts will be out in space again.

I sure hope I will go to the moon some day; I always wanted to go. I was so convinced those craters were made of cheese when I was younger, I wanted a telescope. So my parents bought me a pair of binoculars for Christmas. That was just two years after John Glenn made the rounds. By the way, wasn't he dizzy? Getting back to the binoculars, I couldn't wait until it would get dark Christmas night so I could get a good look at the moon. Don't worry, I was busy all day long looking through those binoculars . . . from everything in the sky to my neighbor's new curtains. I used them to look into peoples' living rooms to see what kind of trees they had. The people across the street had a very little one so I had to go upstairs to get a good look at it.

I was wondering what it is like going to the moon. How long will it be before people can go there? I think it's a fantastic thing that you went! Here, I've never even been up in an airplane! I wish I could buy a piece of the moon. How much does it cost for a little piece of moon dust you brought back?

I kept watching TV all day and night until you finally got out and walked around. I really couldn't believe it. It looked like you were having fun.

When will they have girl astronauts? I have a friend named Diane who would like to go. She'd go anywhere just for the ride. She wanted to go to the moon a long time ago.

I have an uncle who works in the Huntsville Redfield Arsenal as a supervisor. That helped me to get interested in space. Could I please have your autograph? Thank you very much for taking the time to read this.

Sincerely,

Sharon Miletiz
Ottawa, Illinois

Reply from Geneva Barnes

August 5, 1971

Dear Miss Miletiz:

Mr. Armstrong has asked me to thank you for your letter. He is always pleased to hear from those who are interested in the space program and would like very much to reply personally to all. However, I'm sure you understand that this is not possible because of the demands on his time.

In answer to your questions regarding the cost of a lunar rock; these samples are the property of the U.S. Government and are not sold. Understandably, the lunar samples are in short supply and all are used for scientific studies or placed in special exhibitions in museums and planetariums for public viewing.

The fact that we do not presently have a program for female astronauts does not mean that sometime in the future women will not have an active part in manned space flights.

We are enclosing an autographed picture of Mr. Armstrong and some information which we hope you and your friend Diane will enjoy.

Sincerely,

(Mrs.) Geneva Barnes
Secretary to Mr. Armstrong

"PLEASE TAKE ME VERY SERIOUSLY MR. ARMSTRONG"

July 30, 1971

VERY PERSONAL TO:
MR. NEIL A. ARMSTRONG
Former NASA Astronaut

Dear Mr. Armstrong:

I certainly hope that this letter of mine can personally reach the office of the Apollo 11 spacecraft commander MR. NEIL A. ARMSTRONG, who now of course is the Deputy Associate Administrator of aeronautics for NASA here in Washington, D.C. I would be most pleased and extremely grateful if Mr. Armstrong could personally read this letter of mine, but I do indeed hate to bother you sir because of your busy work for NASA Headquarters.

But to get down to my point in writing you this letter sir, please find enclosed with this letter of mine a photograph of the Apollo 11 LM ascent stage being shown as it returns to join the CM. If possible, could you personally autograph this picture for me because it would mean a great deal to me if I could obtain the signature of the first man to set foot on the surface of the moon. Please address it to my name and I would indeed be very honored and most pleased if you could personally do this for me. This is the best photograph that I have available on the Apollo 11 lunar landing mission for you to sign Mr. Armstrong because many of the Apollo 11 pictures are either to dark, or only show astronaut Edward Aldrin on the moon instead of you since you were taking the photographs. But I would certainly be honored if you could personally autograph this photograph for me since you were a marvelous NASA Astronaut assigned as the GT-8 command pilot and especially the Apollo 11 spacecraft commander.

Please take me very seriously Mr. Armstrong and I sincerely wish you all the luck and success to your NASA career with NASA Headquarters assigned as the Deputy Associate Administrator. Thank you so very much sir for your valuable time and most thoughtful consideration to the public; below is my address.

Very sincerely,

Kenneth L. Havekotte
Merritt Island, Florida

Attached to this letter is a note from one of Neil's secretaries that lists all the different requests that Mr. Havecotte had made in writing, all nine of them, dating back just one year to September 30, 1970. Each time Neil's assistants had provided autographs and/or signed photographs. At the bottom of the attached note, in Armstrong's printed handwriting are the underlined words "NO MORE AUTOGRAPHS."

"SO PLEASE SEND SOME GOOD SOUVENIRS"

August 5, 1971

Mr. Neil Armstrong,

I am a student who got interested in space when I had to do a long term science report. I did mine on Apollo 11. Since then I have writen to all of the space instulations I can find. All I ever got was a few little things. I have been trying to get a lot of date, pictures, models & things like that. But recently my arm was broken and I couldn't get money to order from the superintendant of documents, which most all the places refer you to. So since you are an astronaut, who I watched the whole space flight of yours and the rest. So since you can get models and good souvineres easily, I was writing to ask you if you could send me some souvineres. I almost forgot, could you write me a little letter. It would mean so much to me. So please send some good souvineres.

An Interested American,

William Burton
Baltimore, Maryland

"I'M A KID WHO LOVES WATCHING MEN GO TO THE MOON AND BACK"

August 17, 1971

MR. ARMSTRONG,

I would like to ask Some Questions

1. Do you Feel Anything now about Being the First Man on the Moon?

2. ARE you going to make any MORE Space Flights in APOLLO OR SKYLAB?

MR. ARMSTRONG, I think that you are one of the Greatest Man in History. I'm a kid who Loves watching Men go to the moon and Back. I Even Collected a scrap book of all the Missions Beginning with the tragic Apollo 3. I Hope that you go on and do your thing. I Hope Some Day That I Get to meet you. I wanted to go to Wapekeneta But My Dad says no. Neil, Good Luck

Sincerely,

Greg Hartman
Cincinnati, Ohio

"MY SON WAS KILLED IN VIETNAM"

August 30, 1971

Dear Mr. Armstrong,

I have been a very ardent follower of your space flight and have purchased every periodical, book and magazine of your landing on the moon. (In fact I stayed up 38 hrs at one time tape recording all news of your flight.) The one thing I would really like is your autograph.
My only son was killed in VietNam in Oct 1968 and my interest in your flight was the one thing that kept me from really breaking up.
I have arthritis so please excuse writing—please reply

Thank you

Jerry Hammond
Glens Falls, New York

"EVERYONE SAYS YOU CAN'T COME"

September 3, 1971

Dear Sir:

I would like to see if you could come Dinner some time in November. Please everyone sas you can't come. Please do. Write if you would please if you are not you are coming please.

Glenn Smith
Cincinnati, Ohio

"JUST AS YOU MADE A PEACH OF A LANDING"

September 17, 1971

Mr. Armstrong

Dear Sir,

Now getting to the point. So you did on the moon. You see I have been making canes. That I get from the woods. And gave many of them away. With no strings attached. And these folks that I gave them to. Seem to make good use of them. Now I have one that I would like you to have. One that you never have to use. Only as a token and event. Because I was just finishing this cane. When you stepped on the moon. And the top of it is something like a ball. And my wife said that scott of an Armstrong. Just touched the moon.

And I said to her. Yes and I just made my last touch with my brush on said cane. Which is made of Peach-wood. Just as you made a peach of a landing. July 20th 1969. So I would like you to have this walking stick very much. And if you ever make another trip to the moon. Take it with you. You see my grandfather came from Scotland. Alex McEwen.

So you see Neil—I have a wee bit of Scott in me. Now tell me where to send it And will do.

Respectfully yours,

Clifford E. McEwen
"Canes is My Hobby"
Union City, New Jersey

"I HAVE A PROBLEM I THINK YOU CAN HELP ME OUT WITH" (A SERIES OF LETTERS FROM A PERSISTENT MAN IN PENNSYLVANIA)

October 4, 1971

Dear Neil:

I have a problem I think you can help me out with. I have been getting covers autographed by most of the astronauts (Apollo Series) in the hopes of teaching the boy the ways of having a hobby. My problem is this. He says all the covers are autopenned, and signed by a machine as you guys are too busy to sign anything. If this is true I feel like an idiot writing to a machine for an autograph. If at all possible would you autograph the envelope and return it to me as I only have the one cover. I'm going to write to Collins later when I get the cover back. Lastly, congratulations on your new job as professor of engineering at the Univ. of Cincinnati. I wish you all the luck in the world and God Bless You.

Thank You

Charles W. Lentine
Brookhaven, Pennsylvania

P.S. Could you please write on the cover To Charles Jr. then maybe he'll believe its a real signature?

January 30, 1972

Dear Sir:

I know you are a very busy person and probably have very little time for
your own personal business as it is without people writing to you and
bothering you. But I feel I can justify writing this letter to you. Years
back I started collecting covers on space events as they happened, and
then along came my first son after being married for some 10 years. I
promised to myself that I would build him one of the finest space collections that could be built and to give it to him when I felt he was ready
to appreciate it. The boy is getting bigger every day and I feel I am so far
away from completing the covers for him. I had written to many of the
other astronauts and most of them were kind enough to autograph the
covers for the little guy. You would be doing us a very big favor if you
would personally autograph this cover in pen and ink and return it to
the boy in the enclosed envelope.

As the boy gets older he also gets smarter and he questions a lot of
things, and in particular some of the signatures we have acquired. He
says they just don't look real to him. He says you have a machine that
writes autographs by the thousands. I have faith in human nature and
don't believe this. I'd feel like an idiot writing to a machine for its autograph. This is not my way of thinking. I think you guys are the greatest
for helping me out, and also for giving the children of today something
to shoot for in tomorrows world as far as courage, strength and bravery
are concerned.

Maybe someday in the future as the boy looks at these covers with
his children, he'll drink a beer to me for providing him with this cover
collection, and to the space-men for providing the world with the thrills
of the space flights, and for keeping our country number one in the
world. Lastly, congratulations on your major role in the Space Program
and to your courage for being there. May God bless you and watch over
you always.

Thank You,

Gratefully,

Charles W. Lentine
Brookhaven, Pennsylavania

P.S. Could you sign the cover to read TO CHARLES JR. thus personalizing it a bit for the little guy.

February 17, 1972

Dear Mr. Armstrong,

I am not a commercial institution as your secretary referred me as. I am just a small stamp collector trying to do something nice for his little boy. Its a shame you left NASA at least there you had to be nice to the public (those taxpayers) that sent you up to the moon. Now that you are a private individual again you are entitled to a certain amount of privacy. I respect your privacy and probably am the last guy in the world to bother another. I find letter writing very distasteful *and* don't have hardly any time for letter writing as such I am not accustomed to [illegible]ing anyone for their signature. And when the child grows up I will be certain he does not go to the University of Cincinnati where It seems certain individuals get stuck on themselves. I am glad at least some fellows like Shepard, Roosa, McDivitt, Glenn, etc have signed the covers for the boy. Hats off to them. Remember if you want the limelight expect people to bother you.

Sorry to take some much of your time.

Charles W. Lentine
Brookhaven, Pennsylvania

Reply from Geneva Barnes

March 7, 1972

Dear Mr. Lentine:

This is in reply to your most recent letter regarding your request for Mr. Armstrong's autograph on your first day cover. I regret that my February 9 note evoked such a strong letter of criticism of Mr. Armstrong, who has never refused to respond affirmatively to a request for his signature. My records indicate that he signed a cover for you and it was mailed on October 4, 1971.

It would be most unfortunate to disillusion your small son by implying that Mr. Armstrong refuses to acknowledge requests for his autograph and feels that he no longer needs to be considerate of the taxpayers who made possible his trip to the moon. This simply is not true. He continues to honor all requests, which are numerous, for autographs and other information which are directed to him. However, unfortunately for the sincere collector such as yourself, the widespread commercial use of Mr. Armstrong's signature caused him to discontinue signing philatelic items. This policy was put into effect soon after we mailed your first day cover (signed) in October.

We hope Charles, Jr., enjoys the enclosed photograph which Mr. Armstrong has signed for him.

Sincerely,

(Mrs.) Geneva Barnes
Secretary to Mr. Armstrong

August 3, 1974

Dear Prof. Armstrong:

I hope you get this letter in time for our wish of a happy birthday on your birthday of August 5. I know the folks at Wapakoneta, Ohio, must surely be pleased to have you as one of their favorite sons. The way the mails are today I don't know if you'll get this little note on time, but in any case, both me and my young son wish you the happiest of birthdays. I quit counting mine after forty, seems they come to fast. I have a request to ask of you, and I hope it is not too much of a bother. We have been collecting the Smithsonian Institution Set of MILESTONES OF SPACE Covers honoring such great men as: Yeager, Collins, Manke, Van Allen, Wade, MaCready, Rosendahl, Crossfield, Hergett, Hagen, Rickenbacker, Rickover, Lear, Douglas, Loening, Eaker, Gaffaney Hilliard, Von Braun, Webb, Fletcher, Paine, Low, Schirra, etc. and many of the others of Aviation History. This cover released in a series is the 17th and honors your lunar flight aboard the Apollo XI spaceship. If at all possible could you spare us a second to personally autograph this cover and return it in the enclosed return envelope. If you feel this is not possible, just return the cover, and we will ask no more of you. Please one request we make is that no AUTOPEN BE USED, as we have all the other covers in our collection signed personally by these other great men, we would not desire to have people look down on our whole collection, and dishonor thes other men. Thank you kindly.

Gratefully yours,

Charles W. Lentine
Brookhaven, Pennsylvania

PS: Don't eat to much birthday cake, you'll never be able to get off the ground again if you do.

October 25, 1974

To Luanna J. Fisher
Secretary to Prof. Neil A. Armstrong
University of Cincinnati
Cincinnati, Ohio

I hope your University of Cincinnati football team loses all its remaining games for your shabby treatment of my small son. Professor Armstrong is a famous man alrite, but who needs his *AUTOPENNED* signature.

The boy has written to great men such as J. Edgar Hoover, Harry S. Truman, John F. Kennedy, Michael De Bakey, Christian Barnard, Edmund Hillary, Michael Collins, (one of his mates who at least can sit down and answer an honest letter), Gerald Ford, even Spiro Agnew was honest enough to write his own name. Why ruin a boy's cover or Mr. Armstrongs photos with an absurd looking faked autograph. Are people today supposed to be that stupid that they don't even get the courtesy of an honest "no" answer such as men as Charles Lindbergh, Andrew Wyeth, Warren Burger, do. I for me think the Autopen should never be used to fool little guys into thinking that someone was kind to them. Isn't it time we all started back towards honesty—that's whats wrong with our country—nobody has the time or [illegible] to take a second out for the next guy. So be it.

I don't intend to change the world, but here is a guy like Harry Truman who told the truth, no matter what. Please relay this message to your boss, if his head is at least little enough to let his ears let sound in. We apologize to you though Miss Fisher, we realize you are merely a tool & probably have to work for a living just like most of the normal people who pay their taxes for someone to make famous names at their expense. Thank you for lending an ear. God Bless you.

Sincerely,

Charles W. Lentine
Brookhaven, Pennsylvania

P.S. The boy would not dare to show the phoney picture in school—The other kids would laugh at him.

Reply from Luanna J. Fisher, Neil's secretary

October 29, 1974

Dear Mr. Lentine:

How astonished I was to receive your erroneous letter. You are quite mistaken about the autopen. Professor Armstrong takes the time to

PERSONALLY autograph *each* picture request he receives. As explained in the letter sent to you, it is against Professor Armstrong's policy to sign first day covers. That is the reason for the autographed picture as a substitute, and the autograph is GENUINE. If it were not, it would not have been necessary for your son to have waited from August 3, 1974 until October 25, 1974 to receive his signature, as Professor Armstrong was gone for the entire summer and could not personally autograph the picture until this month.

You are right about the need for more honesty—but let me honestly assure you that it is the policy of this office to be honest, sincere and genuine. And I don't consider myself "merely a tool." If I did, I assure you, I would seek employment elsewhere.

Your son need not feel ashamed or embarrassed to display his picture to his classmates or his friends, and he may feel proud in telling them that this is Professor Armstrong's real autograph.

As there was obviously a misunderstanding, I see no need in relaying your message to Professor Armstrong.

Thank you for the opportunity to clear-up your misconception of Professor Armstrong.

Sincerely,

(Miss) Luanna J. Fisher
Secretary to Professor Neil A. Armstrong

"I'M INVITING YOU TO COME HAVE LUNCH WITH US"

October 5, 1971

Dear Mr. Armstrong

Next week is National Hot Lunch Week and I'm inviting you to come and have lunch with us (if you can make it). It would be quite an honor to have you at our school, and besides you're my favorat astronaut. how was it on the moon? It must of been alot of fun. If I had a photogragh of you, I would feel like a million dollars. I've been looking forward to growing up so that I could be an astronaut. I was glued to the TV until the spashdown was over. I even recorded the time and date so I wouldn't forget, what do I do, I go and forget and loose the paper I wrote it on. But if you can't make it just write me a letter and tell me what your going to do. But don't forget try to get here if you can make it. Well I'll be waiting if you can come. Lunch will be at 11:30 Friday October 15, 1971. I hope you can come.

Sincerely Yours

Tom Hickey
Hermiston, Oregon

"THEY ARE TAKING STAR TREK OFF THE AIR"

October 6, 1971

Dear Mr. Armstrong,

Do your children watch "Star Trek" on television? I'll bet they like it. I've decided I want to be an astronaut. There are three things that got me into it. My Father, Nasa space program, and Star Trek. They are taking Star Trek off the air and I thought you could write the network a letter. I'm sure they would listen to you. Please send back a picture (if possible) of your children.

Thanks,

Bob Reese (13 yrs. OLD)
Anaheim, California

P.S. Send the letter to:
Mr. Mort Werner
30 Rockefeller Plaza
New York, N.Y. 10020

"HOW TALL IS THE THING BESIDE THE ROCKET"

October 11, 1971

Dear Neil Armstrong,

Thank you for the information that you sent to me. What are the things like buttons on your space suit for? Could you please send me a closer look of the rockit and Space Control Center. Could you please send me some moon rocks. What equipment did you leave on the moon. How long does it take to build a rocket. What do you eat. What do you do inside of the rocket. What is the number of the controls. How long does it take to reach the moon. How does the ground look. Is it brown or black. How tall is the thing beside the rocket. What is the name of that thing beside the rocket. How many windows are in the rocket. What are the thing in back of the rocket. What is the kind of metal is used to build the rocket. Would you please send me pictures of all of the astronauts. Could you please send me a closer look of the buildings around the rocket. How long does it take to be an astronaut from astronauts school.

Sincerely yours,

Ricky Johnson
Laramie, Wyoming

"I WOULD LIKE TO HAVE A PICTURE TAKEN OF YOU AND ME"

October 12, 1971

Dear Mr. Armstrong,

I am a first year law student here at U.C. When you made your historic trip to the moon however, I was serving as a Peace Corps science teacher in the country of Western Samoa.

In the two years I served, nothing did as much to capture the interest and imagination of my students, and the people in general as the voyage of the Apollo Eleven.

If it could be arranged, I would like to have a picture taken of you and me to send to the newspaper in Samoa. I honestly believe that seeing you with someone they know would be quite a thrill for the Samoans and might serve to raise the status of Peace Corps Volunteers still there. It would certainly be a personal honor for me.

I could take the picture in a few moments with a self timer, and would keep the matter totally confidential here at U.C. The time and place for the picture would be at your convenience. I am willing to set any moment of any day aside to take the photo.

Please rest assured that I will not be offended if you reject my request. I realize that you must have many similar ones.

Yours truly,

Steve Korba
University of Cincinnati
Cincinnati, Ohio

"YOU ARE THEIR HERO"

October 18, 1971

Dear Mr. Armstrong:

This letter is on behalf of the student body of Newport High School, Newport, Kentucky. As a teacher of world history and geography, I have endeavored to build student interest whenever possible. I asked my class what kind of speakers they would be most interested in and the reaction was an overwhelming, "Neil Armstrong!" I assured them that you are a very busy man and that I am no miracle worker, but I would try. Mr. Armstrong, you are their hero. You have been to the moon while a good many of these students have yet to cross the Ohio River to Cincinnati. It would be such a thrill for them if you would come to Newport High

School as a guest speaker. A few of the students wrote letters which I have enclosed.

We realize you are very busy and that your schedule is jam-packed. But we would very much appreciate your visiting Newport for an hour—or a day.

Thank you very much for your time. We'll be looking forward to hearing from you.

Sincerely,

Marilyn Macke
Newport Public High School
Newport, Kentucky

P.S. If you are unable to come, could you suggest a Russian "escapee"? That was my students second choice.

At the bottom of the letter is a handwritten note from Neil: "decline and ask teacher to thank students for their individual letters." The letter that follows was either prepared for Neil's signature or he composed it himself.

Personal reply from Neil

October 21, 1971

Dear Miss Macke:

Thank you for your kind invitation to appear as a guest speaker before the student body of your school. I am sorry that my very heavy schedule makes it impossible to accept your request. My duties here at the university and various government responsibilities limit my outside commitments severely.

I am sorry I cannot suggest a Russian "escapee" either. I do not know any.

Please thank your students for their individual letters.

Sincerely,

Neil A. Armstrong
Professor
Aerospace Engineering

"DO YOU WATCH "I DREAM OF JEANNIE?"

November 30, 1971

Dear Mr. Armstrong:

Hi May I please have your autograph? I have autographs by Al Worden
and Alan B. Sheppard I was wondering if you'd do the same for me!?

   Do you watch "I dream of Jeannie? Its on here every afternoon. I love
it cuz I am a real nut about that kind of stuff.

   I also have 685 autographs and it goes higher and higher all the time.
But sometimes it slows down

Miss Barbara Richardson
Downers Grove, Illinois

"HOW MAGNIFICENT YOU ARE!"

December 8, 1971

MY DEAR MR. ARMSTRONG,

HOW MAGNIFICENT YOU ARE! YOU ARE BOTH A GREAT
ASTRONAUT AND A FINE MAN! YOU HAVE SUCH A VERY
LOVELY FAMILY. YOU ARE HANDSOME BEYOND COMPARE.
YOU ARE A TRUE GENIUS! SUCCESS HAS NOT SPOILED YOU
FOR GENUINENESS AND GOOD QUALITIES MAKE YOU
SUPERB. YOU ARE TODAY'S GREATEST HERO. I AM SURE
YOU'LL BE A GREAT PROFESSOR TOO. MY DREAM-WISH
IS SOMEDAY TO MEET YOU IN PERSON! JUST TO SHAKE

HANDS WITH YOU WOULD BE A GREAT THRILL.

HERE IS WISHING YOU AND YOURS A SPLENDID HOLIDAY SEASON.

WOULD YOU AUTOGRAPH THE ENCLOSED PAPERS FOR ME. IF AT ALL POSSIBLE, I WOULD APPRECIATE AN AUTOGRAPHED PHOTO. THESE WILL BE AMONG MY HEART-FELT TREASURES.

MAY YOUR LIFE BE HAPPY FOR FOREVER.

Very sincerely,

Sally Russell
XXOO
Alamo, Texas

"DID YOU GO ON ANY DIFFERENT PLANET EXCEPT THE MOON?"

February 15, 1972

Dear Neil Armstrong:

I wonder what it is like to be up in space. Isn't it scary? Did you go on any different Planet except the moon?

I think you are very famouse and important.

I wonder if there's another unknown planet up there. I wonder if there's air and water on it.

Isn't it exciting being on the moon?

I'm in fifth grade 11 years old and I love science. My hobby is collecting postcards.

Truly yours,

Sandy Sacher
Stroudsburg, Pennsylvania

"I KNOW YOU BUT YOU DON'T KNOW ME"

March 15, 1972

TO Neil Armstrong

I know you but you don't know me. I'm a big fan of yours. I read and write about you. I am 9 years old. Some day I hope to Be an Astronaut too. It must Be fun to go into space. You are my Favorite Astronaut. How does it feel to go into space? I put Modles together of rockets, planets, and Astronauts. Do you know I have a chemistry set and I have a toy rocket and I am trying to make rocket fuel. If your not to Busy could you send me your picture and autograph it. Some day when your off From work stop over. This is my Address.

Joseph Rosio
Brooklyn, New York

"LIKE TO KNOW WHY PEOPLE ARE SPENDING MONEY ON GOING TO THE MOON"

March 23, 1972

Dear Sir:

I would like to know why people are spending money on going to the moon, when there are people starving. Another question I would like to know is some of the things we have learned from the moon so we would live better. I need to know this information for a classroom project

Sincerely

Brenda Courtney
Sixth Grade
Lincoln Elementary
Gahanna, Ohio

"IF YOU HAD A BICYCLE HORN ON THE MOON . . ."

April 23, 1972

Dear Col. Armstrong,

I have a question to ask you. If you had a bicycle horn on the moon (ha) and honked it would you be able to hear it? Why? I can't write a longer letter because I know your busy and because I have to "hit the hay" as my dad puts it. Well that's all. Good Night!

An interested student,

Dan Schlembach
Toledo, Ohio

Reply from Ruta Bankovskis, Neil's secretary

April 27, 1972

Dear Dan:

Professor Armstrong has asked me to reply to your letter of April 23rd, asking whether a bicycle horn could be heard on the moon. The answer is no. The scientific explanation is that sound is a series of pressure waves (ripples) in the air. Since there is no air on the moon, there is no sound. You probably could feel the vibrations of the horn through your gloves by touching it.

We thank you for writing and hope your curiosity about the universe continues.

Sincerely,
Ruta Bankovskis
Secretary to Professor Neil A. Armstrong

"I HAVE A THEORY"

May 24, 1972

Dear Professor Armstrong:

I am a free-lance writer with a great interest in aviation history and
"old-time" radio adventure programs. I have a theory that many accomplished people involved in aviation today developed an early interest
in aviation because as youngsters they listened to such radio adventure
programs as Captain Midnight, The Air Adventures of Jimmy Allen, Sky
King, Crash Carrigan, Wings of Destiny, and others.

I am in the midst of compiling an article exploring this theory. I am
contacting a number of people such as yourself who have made outstanding contributions to aviation history. I believe most people would
be very interested in your comments on the subject because of the position of esteem you have so richly earned.

I think people would be particularly curious to know which programs
you listened to, which was your favorite, and whether your primary
interest in aviation came about as a result of your early exposure to radio
adventure programs.

Please comment at whatever length you feel appropriate. I will send
you a copy of the article when it is published.

Thank you for your time and trouble.

Sincerely,

William R. Taylor
New Berlin, Wisconsin

Reply from Ruta Bankovskis

June 20, 1972

Dear Mr. Taylor:

Professor Armstrong has asked me to reply to your letter of May 24th asking him to name his favorite childhood radio programs for use in an article on aviation history. We are unable to honor your request as Professor Armstrong does not wish to make public this type of information.

He asked me to thank you for your interest and to extend his best wishes for the success of your work.

Sincerely,

Ruta Bankovskis
Secretary to Professor Neil A. Armstrong

"GOODBYE FOR NOW *HERO!*"

June 8, 1972

Dear Mr. Armstrong,

I want to thank you and Mrs. Barnes for the wonderful pictures and booklets. I will surely show them to the 6th grade.

Someday I hope to meet you in person. And I also accept your apologies. And one thing I want you to remember because I will that you will always be very dear to me though I have not meet you. You see I am the oldest in my family and one thing I always wanted was an older brother and that is how you are treating me. I like it very much learning about space and hope you will teach it to me. Goodbye for now *Hero!*

With all my love,

Sabrina Ezelle
Hattiesburg, Mississippi

P.S. I am giving you a picture of me though I am ugly.
P.S.S. Good-bye *brother!*

"HE HAS LEUKEMIA"

August 26, 1972

Dear Mr. Armstrong,

Several years ago, through Harriet Eisele, my son Todd received an auto-graphed picture of you (which was a great thrill for him & is still a most treasured possession). His birthday party on July 16, 1969 was a "To the Moon" party—& ever after he has felt that day to be doubly important.

He has leukemia & during the past three years has gone through a great deal. Small boys needs heroes—& you are his.

He is not doing well at this time—& the prognosis is not good. If you have a minute from your busy day to drop him a short note—it would mean so much.

Thanks so much in advance—

Most sincerely,

Terry Westhusing
(Mrs. J.K. Westhusing)
for Todd Westhusing (10 years old)
La Porte, Texas

Attached to this letter is a note in Neil's handwriting that reads: "Compose a nice letter to this boy. No mention of his troubles, but about his continuing interest in Space Exploration for my sig."

Personal reply from Neil

September 20, 1972

Dear Todd:

I have recently learned that we both share a common interest—the exploration of outer space. As you may know, my own interest in the subject of aviation began at a very early age and was spurred on with the advent of rocketry and the sudden possibility of actually penetrating the mysteries of the universe beyond our own small planet Earth. The unknown has always held a fascination for me and I consider myself very lucky to have been able to participate in the voyage of discovering another world. I hope that your own enthusiasm will continue as our activities in space exploration progress.

Sincerely,

Neil A. Armstrong
Professor of Aerospace Engineering

Enclosed with the letter was an Apollo 11 patch.

"I ALWAYS LOVED YOU"

September 12, 1972

Dear Mr. Armstrong,

You are my favorite astronaut. I always loved you. I want to ask a favor will you please send me a picture of you and your crew, your rocket ship, splash down and you on the moon. Thank You.

Yours Truly,

Eddie Furman
Brooklyn, New York

"YOU WERE THE FIRST MAN ON THE MOON, CORRECT?"

December 11, 1972

Dear Mr. Armstrong,
You were the first man on the moon, correct? I want some pictures, because I am a big fan of yours!

P.S. Please write back!

Sincerely yours,

Craig Kmiecik
Hillsdale, New Jersey

"DID YOU MEET ANY MARTIANS UP THERE"

January 5, 1973

Dear Mr. Armstrong,
Did you meet any martians up there, you know the moon. Well if you did send me a picture of the strongest one. Did you have a good time. How are you. By the way how's the weather. Well have to go. See ya.

Seeya,

Chris Markus
Williston Park, New York

P. S. You better write back. Because this letter took me a long time to think of.
P. S. S. Write back.
P. S. S. S. Don't forget to write back.
P. S. S. S. S. Goodbye
Turn Over
P. S. S. S. S. S. Send some pictures of the moon. Na forget it.
P. S. S. S. S. S. S. Write Back.
P. S. S. S. S. S. S. S. Seeya
P. S. S. S. S. S. S. S. S. Write Back. Don't forget.
P. S. S. S. S. S. S. S. S. S. Goodye This is last P. S. or P. S. S. S. S.

"SO DO YOU GET HICCUPS IN SPACE?"

January 15, 1973

Dear Mr. Armstrong

I am a great admirer of yours. My friend and I are building a rocket, I wonder if you can give some tips. Our rocket is going to be 3 feet high and ½ a foot wide. This question may seem strange, but I am asking this because of no gravity and all the pressure against you, so do you get hiccups in space? Please reply. Thanks for reading.

Sincerely,

Gary Jones
Henry W. Longfellow School
Pasadena, California

"I AM WRITING A BOOK ABOUT YOU"

March 6, 1973

Dear Commander Armstrong,

I am 8 years old. My name is Howard Levy. I have watched the rocket lift offs and splash-downs since your landing on the moon. I would like to be an astronaut when I grow up. I am writing a book about you.

I would like to know how you became an astronaut. I will be glad to send a copy of the book if you would like one. Thank you for helping me with my book.

Sincirely

Howard Levy
Blackstone, Virginia

"FROM NOT *THE* RICHARD NIXON"

July 21, 1973

Dear Sir

As the *First* man to set foot on the Moon, I guess you should be the first Astronaut to be on My Distinguished List of great and famous people's personally signed letter. Size about 10″ x 12″ on your finest astronauts stationary—Picture size should be about 8″ x 10″. The main reason I mention size is because each personally signed letter or personally autographed picture and in your case I would sure like to have both, colored picture if possible.

You might say this guy sure has a lot of nerve to ask for a favor and then ask for an almost exact size. This is because each letter and/or picture or both are placed on a separate page, under acetate cover, and put in a very exsspensive Leather Binder—Size about 11½″ x 12″ also I might add that out of 300 requests, I have only *one* or *two* refushed to answer my letter. I am working on VOL IV at the present time and this is over a period of two years.

The following zerox copies will give you a good idea of my success in this very interesting, rewarding—different, unique, and valuable collection. In fact it is fast becoming (one of a kind) in the entire country. You probably won't believe this but $25,000 would not buy it from me at the present time and in the years ahead (not in my time or yours) it will be priceless, as you already know their are some things (even Howard Hughes could not buy).

Just a short history of myself. I was born in Mansfield, Ohio 3/16/13 and after graduation from High School, I went to work for the Great Prudential Jus. Co. of America as an agent in my own home town. After serving the company in various catagories for 32½ years of loyal service—the pressures became too great for me. So on 2/27/68 I was forced to go on Disability Retirement. Consequently, one might say that after moving to Las Vegas, Nev. My doctor said I needed a real good Hobby. After careful consideration this *one* was right down my alley (probably from my great success in the Insurance Field).

Thanking you in advance for reading this lengthy epistle, I will be awaiting your early reply.

Sincerely

Richard W. Nixon
Las Vegas, Nevada

P.S. Please let me know if I can reach all the Famous Astronauts at this same address.

Attached to the letter that Mr. Nixon sent to Neil a list of over 100 celebrities for which he claimed to have "personally signed and authenticated letters"—a list that included President Richard M. Nixon, his wife Patrician Nixon, and the Nixons' two daughters, Julia Nixon Eisenhower and Tricia Nixon Cox. He also sent a business card showing the Nixon family crest. If Neil or his secretaries responded to Mr. Nixon's letter, I have not found that reply.

"THE LIBRARY OF CONGRESS COULD NOT ADEQUATELY RESPOND"

September 11, 1974

Dear Professor Armstrong:

This is an attempt to gain clarification of the remarks attributed to you upon your landing on the moon on July 20, 1969. It is also a request upon your busy schedule to help us correctly comprehend the true significance of your thoughts.

The Library of Congress could not adequately respond to our inquiry, and referred us directly to you.

It is recorded that upon your first step upon the moon you said:

"It is one small step for (a) man,
one giant leap for mankind."

The emphasis of distinction is whether the comments you made were, ". . . step for man . . ." or ". . . step for *a* man . . ."

Until recently we were of the opinion that you represented all of mankind with your first step and accepted the responsibility without being overcome with the awesome nature of the event. With the latter emphasis it seems to indicate that you recognized your own fallibility as a man who seemed to comprehend his place in the cosmos.

We tend to identify with the latter as we all are but a part of humanity working for mankind. Consequently we ask that you, if you would, help us comprehend a bit of history more clearly. The moon landing holds a special significance in our life. The fact that it was you who made the first step that help clarify the misunderstanding, would be greatly appreciated.

Respectfully,

Rody and Doris Yezman
Alexandria, Virginia

Reply from Luanna J. Fisher

October 3, 1974

Dear Mr. & Mrs. Yezman:

Professor Armstrong's statement, intended to include the (a), was simply intended to mean that, although the step from the lunar module to the surface of the moon was simple for a man, the importance, philosophically, for the future of humankind was enormous.

I hope this is of help. Professor Armstrong appreciates your interest.

Sincerely,

(Miss) Luanna J. Fisher
Secretary to Professor Neil A. Armstrong

"SOME OF THE KIDS DIDN'T BELIEVE THAT
YOU HAVE BEEN TO THE MOON."

December 6, 1974

Professor Armstrong,

I'm a 9th grade student at Edmonson County High School.

One day while we were studying science the subject came up about the Astronauts. Some of the kids didn't believe that you have been to the Moon. But myself I do. So could you send me some kind of information to take to school that will prove that you went.

My teacher Coach Dave Denton tried to explain but they still couldn't understand.

Thank You,

Rebecca Ann Decker
c/o Mr. Leon Decker
Bee Spring, Kentucky

On this letter Neil made a note to his secretary Luanna J. Fisher: "Lu—Tell her it's not necessary to prove it. When they are a bit older they'll believe. N"

Reply from Luanna J. Fisher

January 10, 1975

Dear Rebecca:

Professor Armstrong has asked me to thank you for your kind letter. He says to tell you that it is not necessary to prove that he went to the moon. When the students are a bit older they'll believe.

Thank you for your interest.

Sincerely,

(Miss) Luanna J. Fisher
Secretary to Professor Neil A. Armstrong

"WOULD YOU HELP US BY LISTING SEVERAL BOOKS YOU HAVE ENJOYED"

September 22, 1975

Dear Mr. Armstrong:

Our library is planning a display using the theme of favorite books. We are writing to various well-known people to ask for their suggestions. Would you help us by listing several books you have enjoyed and would like to recommend to others? We will display your response to our college students, faculty and the general public.
Thank you.

Yours truly,

Terry Ann Forster
Reference Librarian
Lane Community College
Eugene, Oregon

At the bottom of Ms. Forster's letter, Neil printed the names of the following books and authors:

> The Papers of Wilbur & Orville Wright, *McFarland*
> Fate is the Hunter, *Ernest Gann*
> Red Giants & White Dwarfs, *Robert Jastrow*

I did not find a reply to Ms. Forster but assume that one of Neil's secretaries prepared and sent a letter to the librarian giving his short list of recommended books.

"THEY WILL CHERISH THESE BOOKS
FOR THE REST OF THEIR LIVES"

June 8, 1979

Dear Mr. Armstrong:

I want to shoot the moon and ask you for a small favor; in fact, two small favors. Let me explain.

Ten years ago I took my two young daughters (Francie, age 13 and Kate, age 10) down to Cape Kennedy to watch your momentous lift-off to the moon. I had each of the girls keep a log of extraordinary events before and after the lift-off. Each day into their logs they wrote their own observations along with newspaper accounts, photographs, maps, etc. which they pasted down on separate sheets.

I recently came upon these two logs and I am about to have them bound in covers to present to each of my daughters on the tenth anniversary of Apollo XI. The thought came to me that the final page in these two logs could ideally be a short note from you to each of the girls. I have no specific thoughts to suggest on this matter with the exception to wish them well on their own future endeavors (Francie, now age 23, just graduated last week with honors from the University of New Hampshire and Kate, now age 20, is a junior at Northwestern).

I know that both of the girls have long since forgotten about the logs they kept. Once they see them again, I know they will cherish these books for the rest of their lives.

If you can supply me with two notes and send them to the above address, I will be extremely grateful.

Cordially,

Nicholas Benton
Vice President
Time-Life Books Inc.
Alexandria, Virginia

P. S. I enclose copies of the July 16, 1969 pages from their logs for your enjoyment.

Francie Benton's account

July 16, 1969

Today we stumbled and groaned out of bed at 2:00 in the morning. Hoped into the Avis car that we rented. We expected the roads to be jam packed, but the highway was free of cars all the way to Coacoa Beach. In Coacoa Beach we had breakfast in a small coffee shop. It took us about an hour to find Colberts Marina, and by then it was 5:45 and the traffic started piling in. At about 7:00 the boat started down the Indian River toward Cape Kennedy. The boat ancored 3½ miles away.

We heard the count down on the radio, at 1 second fire started spitting out from beneath the rocket. We couldn't hear any noise untill the rocket was about half a mile into the air and then it sounded like machine guns. I cried a little thinking that man was really going to the moon.

Kate Benton's account

Wed. July 16, 1969.

At two o'clock in the morning Dad woke up Francie and me. We got dressed quickly and got in the car to drive from our hotel in Daytona to Cape Kennedy. Francie and I fell asleep in the car but we woke up when Dad pulled over to the side of the road. He told us to look out the window and there on the horizon was the rocket.

It was beautiful! The search lights were lighting it up and it was bright white. The rocket was or looked one foot tall but it is really much bigger.

After we had some breakfast we drove to Mr. King's boat (a friend of Dad's). It was now six o'clock and the dawn was coming up. We started off at seven and went up the Indian River where we saw some dolphins. Mr. King anchored his boat three and a half miles away from the launching pad.

We listened to the countdown on the radio and I looked through the binoculars and all of a sudden the rocket went off! It looked really great when the flame let off all the smoke. As it went off it felt like there was a strong thunder shaking the world.

It was so fantastic! and very bright. When it went out of sight it burned a hole through the cloud and then it was out of sight.

After the lift off Mr. King served champagne so we could salute the three astronauts.

Personal reply and letters from Neil

October 2, 1979

Dear Mr. Benton:

While working my way through the stacks of mail that had accumulated during a busy summer away from the campus, I came across your letter and the Apollo launch accounts of your daughters, which I enjoyed thoroughly.

The notes for your use are several months late, but you may still want them, and so they are enclosed.

In any case, please give my very best to Francie and Kate.

Sincerely,

Neil A. Armstrong
Associate Director
Institute for Applied Interdisciplinary Research
University of Cincinnati

October 2, 1979

Dear Kate:

Ten years ago, we shared an experience: you from one vantage point, I from another. I hope you will remember the occasion with some warmth and satisfaction, as I do.

I enclose my sincere good wishes for your future happiness.

Sincerely,

Neil A. Armstrong
Associate Director
Institute for Applied Interdisciplinary Research
University of Cincinnati

October 2, 1979

Dear Francie:

It is impertinent for me to address you so personally, but it seems appropriate. I know you through your account of an occasion we shared ten years ago; an occasion I remember clearly.

Over the past decade, you have grown to adulthood and have excelled in what we call "higher education." I can only hope that you will enjoy a life filled with events that you consider worthy of being categorized as "human progress."

Sincerely,

Neil A. Armstrong
Associate Director
Institute for Applied Interdisciplinary Research
University of Cincinnati

"I WAS JUST WONDERING WHERE YOU THOUGHT IT UP"

October 25, 1980

Dear Mr. Armstrong,

You are, of course, a great person, and I am very proud to be able to write to you.

The quote you made when you first stepped on the moon ("It's one small step for man, one giant leap for mankind") is a quote that will live for ever in men's hearts. I am only 14 and don't remember when man first set foot on the moon, but that quote has been with me for as long as I can remember. It is one of my favorite quotes (besides Bible verses), and I was wondering how it ever got started. I know you probably knew you would have to say something great when you first stepped on the moon, and I'm sure you knew it would become famous. I was just wondering where you thought it up. Whether it was on Earth, on the way to the moon, or if it just popped into your head right before you stepped on the moon.

I would love it so much if you could answer, and I have enclosed a self-addressed stamped envelope so it won't cost you anything to reply. Thank you so much.

Very sincerely,

Richard Lawson
Madera, California

Reply from Vivian White, Neil's assistant

November 7, 1980

Dear Mr. Lawson:

Mr. Armstrong has asked me to reply to your letter and to thank you for your interest.

Mr. Armstrong conceived the statement after landing on the moon, but prior to emerging from the craft.

Sincerely,

Vivian White
Assistant to Neil A. Armstrong

6

RELUCTANTLY FAMOUS

The high price of fame was a heavy burden for all the early U.S. astronauts, but none paid more dearly than the First Man on the Moon—as personally unwanted as his status as a celebrity and global icon immediately became after he made that "one small step" onto the Sea of Tranquility on July 20, 1969. It was an inevitable legacy that Neil Armstrong never asked for, unhesitatingly did not want, and, for his entire post–Apollo 11 life, had an unusually hard time bearing.

Neil absolutely hated being famous—from start to finish, no doubt about it. He hated it not because as the first man on the Moon he had become an immortal figure forever more in the history books—he was immensely proud of the Apollo 11 mission, he understood what great sacrifice, what awesome commitment, and what extraordinary human creativity it had taken to get that first lunar landing mission done and brought back home safely.

But Neil also knew, and always emphasized whenever the subject came up for discussion (which for Neil was always too much), that it was the teamwork of some 400,000 Americans that enabled Apollo's success. As the commander of Apollo 11, he had been at the top of that pyramid—the foremost member of the extraordinary team that achieved humankind's epochal first step onto another heavenly body. But, as the global public did not clearly recognize at the time in 1969—nor does it today—that there had been nothing preordained in his becoming the commander of the first Moon landing or, furthermore, becoming the first man out onto the lunar surface. As Neil always tried hard to explain, those tasks were mostly the luck of the draw, emerging from an unpredictable complex of highly contingent circumstances. As he explained to CBS journalist Ed Bradley for a *60 Minutes* profile in November 2005, the "principle concern" of Deke Slayton, the head of the astronaut corps and the man who put the Apollo crews together, was "getting a qualified capable commander on each flight." Deke "took the position that the guys had all come through the process, were all qualified to fly, should be able to fly, and should be able be able to accept any task they were given." So, "I wasn't chosen to be first. I was just chosen to command that flight. Circumstance put me in that particular role."[20]

But the public did not want to hear such pedestrian explanations from Neil or anyone else; it wanted heroes and heroic tales, especially with its astronauts—the First Man foremost among them. Neil hated the iconography that came to surround and define him. He did his best to correct and deflect the epic and monumental elements that society and culture built into his legacy, which he knew greatly distorted who he actually was, led to countless myths, errors of fact, exaggerations, apocryphal stories, and downright lies about him, what he had done, what he believed, what he stood for. He hated fame and celebrity for its constant intrusion into his personal and professional life—and most of all, into his daily, even minute-by-minute, thoughts, into his very living consciousness. In the extraordinarily modest, unassuming, and private way he lived his life after Apollo 11, it was clear that Neil understood that the glorious feat that he helped achieve for the country back in the summer of 1969—glorious for the entire planet—would inexorably be diminished by the blatant commercialism, redundant questions, and noise of the modern world. Whether it was some nobility in his character or something less noble, Neil just would not let himself take part in any of that. He was a man who just could not be bought.

Almost everything about being the First Man on the Moon took a heavy toll on him—and nowhere was the toll on Neil more sorely affecting than in his voluminous public mail and his laborious efforts to handle it over the course of the forty-three years he lived after Apollo 11. The barrage was never-ending. Although it never again reached the rate of the some 10,000 pieces of mail per day that he received in the weeks and months following the Moon landing, the pressure of answering to the public never much receded. Without question, Armstrong's signature, through the entire history of the Space Age from 1969 to his death in 2012, remained by far the most popular sought-after astronaut autograph. Enthusiasts called it "the holy grail" of autographs. Handwriting experts published articles on how the downstrokes to his "N" and "A" and other features of his signature changed over the years. Though not rare—given how many signatures Neil had provided freely for the public for over twenty years, and given the perceived value and the desire to possess his autograph—demand always remained high. After his death, that demand increased greatly, as did the price for any Armstrong autograph or item of memorabilia. (Neil's death in August 2012 brought more examples of his autograph to the market, and the same happened in 2019 because of the fiftieth anniversary of the Apollo 11 mission. As a result, serious space memorabilia collectors have become more selective in their purchases, placing a higher importance on the provenance of each piece, where, at the same time, highly publicized auctions have drawn new collectors and speculative investors who are not familiar with the market and who have paid more than perhaps the auto-graph can maintain.[21])

As the letters in this book make so disconcertingly clear, people wanted so much from Neil, more than he could ever give. But he tried. It was very hard for him, but he tried. So many requests came in to him asking for so much "stuff"—autographs, pictures, brochures, books, information, answers—not just from children but also from many adults. Most requests were polite but some were not. Many fans were writing, they said, to get Neil's autograph and then gift it to a child, a grandchild, a father, a friend, or someone who was very ill or dying. Some of those stories were no doubt true, but many of them were not.

It took a lot of time and energy for Neil's fans to write their letters. For many it was probably the first fan letter they ever wrote. It was a spe-cial thing for them to be writing to their great hero. They asked a lot of good questions but many of the questions could have easily been found in books or encyclopedia articles. But the fans wanted to hear the answers directly from Neil, who, in the minds of many of them, seemed to possess the secrets of the universe. Some of the letter-writers sought his actual companionship and a few even his love. As we have already seen, several teenage age girls wrote long love letters to him, a few on a weekly basis, with enclosed pictures and scented hankies.

None of this was easy for a man like Neil Armstrong, but one can spec-ulate that no other astronaut could have done a better job with it. Neil was a very modest man, and his modesty shows up in most of his letters. Many of his replies were tremendously witty and insightful—many also were sharply honest and some even sarcastic. Reading through all of the letters put together for this chapter, one could easily come away not liking Neil very much, thinking him too strict, too ungenerous, too principled, not just for his fans but for his own good as well.

Only the people closest to Neil truly saw the effects on him. His first wife, Janet, who quietly divorced him in 1994 because, in her words, she "just couldn't live with the personality anymore," knew better than anyone what it had all done to him inside. "He feels guilty that he got all the acclaim for an effort of tens of thousands of people. Neil would let it bother him. He always was afraid of making a social mistake," though he had "no reason to feel that way for he was always a well-mannered gentleman." He "certainly led an interesting life, but he took it too seriously to heart. He didn't like being singled out or to feel that people were still wanting to touch him or get his autograph. Yet he wouldn't quit signing autographs for twenty years because probably, in the bottom of his heart, he didn't think most people were trying to make money selling them." In Janet's view, if he had gone out in public more times over the years and shared more himself in that way, the more obsessive interests in him would have dwindled; instead, he made himself a type of target.[22]

Could another human being have handled the bright glare of international fame and instant transformation into a global icon better than Neil did? It depends on what one means by "better." Many individuals, without question, would have found ways to benefit more materially—and done less suffering—from the fame than Neil did.

But it is hard to imagine anyone who worked harder to remain who he was than Neil Armstrong—or who experienced more pressure from the outside world to change his character. "The privilege of a lifetime is being who you are," wrote mythologist Joseph Campbell in *Reflections on the Art of Living* (1991). As extraordinarily hard as it was for such a famous man to do, Neil exercised that privilege every day of his life—and all of us who look back on his remarkable life should be happy that he did.

(Readers should note that many of the letters in this chapter are from Neil in response to an initial request letter that I have been unable to locate.)

"IT WOULD BE NICE TO HAVE YOUR PICTURE ON DISPLAY"

[Translated from Italian at NASA Headquarters]

August 21, 1970

Dear Neil Armstrong:
I run a tailor shop here that has the name "Neil Tailors." Since the name is the same, I thought it would be nice to have your picture to display in the store. Can you send me one? I will be most grateful."

Mrs. Milena Rampinelli
Bergamo, Italy

"*THIS GUY* SENDS A LOT OF AUTOGRAPH REQUESTS"
(A SERIES OF REQUESTS FROM A MAN IN NEW YORK)

September 14, 1970

Dear Astronaut Armstrong,

Enclosed is a copy of the book "First on the Moon."
    Needless to say this was a most interesting and exciting story of this historical event in our time.
    I keep books that I enjoy reading again in my small library.
    Would you be kind enough to inscribe this book for me and I will cherish even more within my library.
    Thank you for this consideration.

Respectfully,

Louis Newman
Liberty, New York

March 12, 1971

Dear Commander Armstrong,

Enclosed is a copy of the book "Exploring the Moon." This book has already been inscribed by Astronaut Edwin Aldrin—
    With the true desire to perpetuate the history of Man's Flight to the Moon—I read this book and enjoyed this immensely.
    I would like to keep this with my small library of books that I enjoy reading over again—
    If you would be kind enough to inscribe this book for me, I truly would be honored to include this in my small library of cherished books dedicated to Apollo—
    Thank you for this consideration.

Respectfully,

Louis Newman
Liberty, New York

Handwritten by Neil at the bottom of this letter: "I signed the book."

Note to Neil from his secretary, Geneva Barnes

June 12, 1971

Mr. Armstrong:

See your note re Mr. Newman. According to previous correspondence, he has received an autographed first day cover, and two books.
    He has now sent a small print which has been signed by 6 of the original 7 astronauts and he is asking that you sign it. In addition, he asks "What are your thoughts to Apollo 11 and the future of our space station during our current space austerity program . . ."
    I have prepared a response to his question using quotes from your War College and West Point speeches. But it occurred to me that you may not wish to sign the poster, in view of your note.

Gennie

Handwritten at the bottom of this note and signed by Neil: "THIS GUY sends a lot of autograph requests. N.A."

June 19, 1971

Dear Mr. Armstrong,

With the sincere desire to perpetuate the history of Pres. John F. Kennedy, I am compiling A Album dedicated to Pres. Kennedy and his association with the Space Landing on the Moon.

Enclosed is a beautiful association print which has already been sent to and signed by Astronauts Shepard, Schirra, Glenn, Cooper, Carpenter and Slayton—If you would be kind enough to inscribe "One Small step for man, one Giant Step for Mankind" and autograph this for me, I truly would be honored to have this mounted at the head of my dedication to Pres. John F. Kennedy.

Mr. Armstrong, what are your thoughts to Apollo 11 and the future of our Space Station during our current Space Austerity Program and what can I do to help bring back our full interest in the United States Space Program.

Most Respectfully,

Louis Newman
Liberty, New York

Reply from Geneva Barnes

June 30, 1971

Mr. Louis Newman
Liberty, New York

Dear Mr. Newman,

Mr. Armstrong has asked me to thank you for your recent letter. He would like very much to reply personally to all of the mail he receives but this is not possible because of the demands on his time.

He has signed the print for you but did not inscribe his famous quote, as you requested. Mr. Armstrong does not quote himself.

Sincerely,

(Mrs.) Geneva Barnes
Secretary to Mr. Armstrong

October 27, 1971

Astronaut Armstrong,

With the true desire to perpetuate the history of "Man's Flight thru Space," I am compiling A Album dedicated to this project.

Recently, I was fortunate to obtain this beautiful card honoring A Decade of Space Achievements—.

If you would be kind enough to inscribe this card with a thought associated with Man's Flight thru Space—I truly would be honored to mount this at the head of my Album.

I would appreciate any information, photos or possibly a note reflecting your thoughts on the critics of our Space Program. I'm constantly defending this project and could use more information.

Thank you for this consideration.

Respectfully

Lou Newman
Liberty, New York

Reply from Geneva Barnes

November 15, 1971

Mr. Louis Newman
Liberty, New York 12754

Dear Mr. Newman:

We are returning your first day covers unsigned. Mr. Armstrong has discontinued signing all philatelic items because of their widespread commercial use.

I apologize for the delay in returning your covers. However, our volume of mail has been such that we have not been able to reply as quickly as we would like.

Sincerely,

(Mrs.) Geneva Barnes
Secretary to Mr. Armstrong

March 27, 1972

Dear Astronaut Armstrong

As it is generally known in Liberty, N.Y. that I am a Ardent Supporter of the U.S. Space Program and a even stronger admirer of Astronaut Neil A Armstrong, I was recently fortunate to obtain this beautiful enclosed photo—

If you would be kind enough to inscribe this photo for me with a thought associated to Apollo 11—I truly would be honored to mount this within my Album of Tribute to Astronaut Neil Armstrong and the U.S. Space Program—

Thank you for this consideration

Most Respectfully,

Louis Newman
Liberty, New York

Reply from Fern Lee Pickens, Neil's assistant

July 5, 1972

Mr. Louis Newman
Liberty, New York

Dear Mr. Newman:

Mr. Armstrong has asked me to reply to your letter of March 27, 1972. I apologize for the delay in answering, but we receive thousands of letters from all over the world and our volume of mail is so heavy that we are unable to make prompt reply.

In this your most recent letter, you ask for another autograph. Our files show that Mr. Armstrong has been very generous in autographing items for you, having supplied you with five between October 1970 and December 1971. I am sorry, but we cannot honor your further requests for autographs and your picture is returned herewith.

We are enclosing some information which we hope you will enjoy; also, a brochure of NASA Educational Publications showing the kinds of material available and where to obtain it, should you be interested in adding to your personal library and your albums.

Your interest and enthusiasm in the space program is appreciated and we hope it will continue as our activities in space progress.

Sincerely,

(Mrs.) Fern Lee Pickens
Assistant to Mr. Armstrong

"A PERSONALLY SIGNED CHRISTMAS CARD"

January 14, 1971

Mr. Selwyn C. Gamble
Senatobia, Mississippi

Dear Mr. Gamble:

Please accept our apologies for the delay in responding to your mimeo-graphed letter of November 25 in which you requested a personally signed Christmas card and another personal item from Mr. Armstrong. Because of the incorrect address and the heavy holiday mail, your form letter did not reach Mr. Armstrong's Washington office until after Christmas.

Unfortunately, because of his heavy schedule of activities with NASA and the tremendous volume of mail he receives daily, Mr. Armstrong is unable to personally correspond with his many admirers throughout the world. We are, however, enclosing an autographed picture of Mr. Armstrong which we hope you will enjoy adding to your collection of memorabilia.

Best wishes and, again, thank you for your interest in writing.

Sincerely,

S. B. Weber
Assistant to Mr. Armstrong

"I AM RETURNING THE SHEETS UNSIGNED"

March 5, 1971

Mr. Max J. Lewallen
Northeast Chapter, GSPE
Southern Piedmont Conservation Research Center
Watkinsville, Georgia

Dear Mr. Lewallen:

I am replying on behalf of Mr. Neil Armstrong to your letter of February 18 requesting that he sign the enclosed stationary. I am returning the sheets unsigned as Mr. Armstrong does not approve of his signature being used for fund-raising activities, as a prize, or as a gift and never knowingly supplies an autograph for any of these purposes.

We certainly appreciate your interest and extend our best wishes for much success in your endeavors.

Sincerely,

S. B. Weber
Assistant to Mr. Armstrong

"YOUR FOOT IS PLACED IN A NOTED AREA IN MY OFFICE"

March 5, 1971

Dear Neil,

As I did not get the opportunity of meeting you before your departure, I am writing to ask a favor.

As the Foot Specialist of the NASA Area, your foot is placed in a noted area in my office. It is the picture of your foot and footprint on the moon.

Your signature would be greatly revered and appreciated. May I send you a photograph for your hand? I did not want to send it without your permission.

Your consideration would be greatly appreciated!

Very sincerely yours,

Gilmore E. Guster, D. P. M.
Webster, Texas

The famous photograph taken on the lunar surface during Apollo 11 to which Mr. Guster refers captures the footprint of Buzz Aldrin, not Neil Armstrong. To this day, many people think they are seeing Neil's footprint on the Moon when they see this photograph. No photo of Neil's lunar footprint was taken.

"14 SIGNATURES REQUESTED"

May 18, 1971

Mr. Armstrong:

Brent Mulcahy brought this in. There are 14 signatures requested; all by Congressman Whalen's office.

1. autograph for picture of you and the Congressman
2. autograph for picture of you and the Congressman's Executive Assistant, Bill Steponkus
3. 10 lithos to be used in London, Berlin and Augsburg for hosts of officials of Dayton on their Sister City tour
4. Reproduction from the Dayton Daily News on July 21, 1969
5. First Day of Issue poster containing President's and PMG's [Postmaster General's] signature

There's a ltr attached from the Congressman to Dale Grubb.

Gennie

At the end of her note to Neil, Geneva Barnes added in her handwriting: "The Congressman is pushing for these as his constituents are leaving soon on the 'Sister City Tour.'"

Charles W. Whalen Jr. (1920–2011) represented the Ohio 3rd District in the U.S. House of Representatives from 1967 to 1979. Prior to that Whalen served in the Ohio House of Representatives (1955–1961) and the Ohio Senate (1961–1967).

"COULD YOU SEND ME ONE OF YOUR SOCKS"

August 5, 1971

Dear Mr. Armstrong,

My name is Fiorella. I am 13 years old and collect socks of the "Greats of the World."

Could you send me one of your socks, with a certification that you actually wore it. Could you also send me a photograph of yourself.

Sincerely,

Fiorella Giusti
Sermide, Italy

"TWO MINUTES OF YOUR VALUED TIME"

August 30, 1971

Dear Mr. Armstrong:

After learning through our local news that the famous "First Man on the Moon" is about to become a Cincinnatian, a million and a half people, like myself, are filled with pride beyond belief.

I am sure, every human soul in the Greater Cincinnati Area joins me in extending a warm welcome and a successful stay in the Queen City to "Our Man Neil."

Being mindful of your very busy schedule, I would like to state a request on behalf of an eighteen year old German boy, who is about to take his first trip across the ocean to visit our great nation. It is difficult for me to describe adequately this boy's admiration for you. To say Hello and shake your hand would turn his visit into an adventure beyond comprehension.

He will be in Cincinnati from October 11, 1971 through December 14, 1971 as a guest in our home. We would be more than pleased to take him anywhere convenient to your schedule and two minutes of your valued time would mean a life time of excitement for him.

Since no one will be at our residence from October 3rd to the 14th, a collect call before October 3rd would be gratefully accepted.

Respectfully yours,

Horst Hickman
Cincinnati, Ohio

"NOT BE MY INTENTION TO BE INVOLVED IN THIS KIND OF ACTIVITY"

October 14, 1971

Mr. James W. Hancock
Lucas County Republican Party
Toledo, Ohio

Dear Mr. Hancock:

Thank you for your letter and invitation to participate in the Lucas County Republican Party Rally.

I will, on those dates, still officially be a government employee and hence may not legally participate. However, even if this were not true, it would not be my intention to be involved in this kind of activity.

Thank you for your inquiry and interest.

Sincerely,

Neil A. Armstrong
Professor of Aerospace Engineering

"I AM EXPECTING HELP FROM YOU"

October 29, 1971

Dear Sir,

I wish you good luck. God bless you! I am an Indian student. I like you very much since you are the first man who set foot on the moon. This is really a great adventure for, we human beings. And by this adventure you have obtained a good name from us.

Sir, I am interested in science. But, as I am being poor, I have no facilities to study. If you have pity on me, I can study. I am studying in the pre-university class in our college. I am unable even to pay the fees to the college. I have no mother. During my childhood, she had passed away. My father is an old man, refuses to give money for me. My brothers are in good positions. But they too do the same. In this situation, I am expecting help from you. If you help me, now, I will not forget you for ever. Please help me only for studying purpose. I am just weaping on seeing my posture.

If you want to help further, please send me books about ZOOLOGY and ASTRONOMY. I am interested very much in them. Can you send me man landing on the moon—photos? I want to receive a long letter from you and it should be written in your own hand. It must contain your feelings when you were away from the earth, till you return to the earth. If you send me a letter written thus by your hand, I will think that this is the use of my birth.

I humbly request you to send me your separate photo and your family photo. I hope you will send me, all of my requists.

Thanking you,

Yours sincearly,

R. Sigamani
Madras, India

Reply from Geneva Barnes

May 23, 1972

Mr. R. Sigamani
Madras, India

Dear Mr. Sigamani:

Mr. Armstrong has asked me to reply to your letter of October 29.
I apologize for the delayed response. We regret that Mr. Armstrong
cannot provide monetary assistance to help you with your studies. It is
financially impossible for him to honor requests of this nature as he has
received so many. Perhaps you could contact a local city official or minister for advice on the best source for assistance.

Sincerely,

(Mrs.) Geneva Barnes
Secretary to Mr. Armstrong

"LIKE TO KNOW YOUR SHOE SIZE"

October 29, 1971

Dear Mr. Armstrong,

ZIP Line is a public service feature of the [Toledo, Ohio] *Blade* which
tries to answer questions and help solve problems for our readers.

We have a question from four young readers who would like to know
your shoe size. While camping under the starts—and moon—they say
they had a discussion about what size shoe you wore. They think they
saw the foot print.

If you don't mind providing the information we'd appreciate it and pass it along to our readers. Also if you have some idea of the length and width of the print your space shoe made we'd like to include that too.

Thanks for your time.

Yours,

Barry Stephen
ZIP Line
Toledo Blade

Reply from S. B. Weber, Neil's assistant

December 18, 1970

Mr. Barry Stephen
ZIP Line
The Blade
Toledo, Ohio

Dear Mr. Stephen:

Thank you for your letter of October 29, and please accept our apologies for the delayed response.

In answer to your question, Mr. Armstrong wears a size 9 shoe. We are happy to have been able to provide you with this information to pass along to your readers.

With kind regards,

Sincerely,

S. B. Weber
Assistant to Mr. Armstrong

"INAPPROPRIATE TO PARTICIPATE IN A DEDICATION OF A BUILDING BEARING MY NAME"

November 11, 1971

Mrs. Robert Lieber
Neenah, Wisconsin

Dear Mrs. Lieber:
Thank you for the kind letter and request to reconsider my answer to Dr. Scott's invitation to join in the dedication. As I stated in my letter to Dr. Scott, I feel it is inappropriate to participate in a dedication of a building bearing my name. I have maintained a consistent policy in this regard for a number of years.

Thank you very much for your interest and best wishes for the success of dedication and your new school.

Sincerely,

Neil A. Armstrong
Professor of Aerospace Engineering

Dr. Donald Scott was the superintendent of Neenah High School. The high school is comprised of three buildings, one of which was named after Neil Armstrong in 1972.

"I HAVE BEEN APPROACHED BY A NUMBER OF AUTHORS"

January 12, 1972

Mr. Jeremy M. Harris
Worthington, Ohio

Dear Mr. Harris:

Thank you for your letter regarding a possible biography directed toward the junior high school and high school levels. I have been approached by a number of authors who have considered similar projects.

I would not be interested, at the present time, in collaborating on a biographical text of any kind, nor am I inclined to endorse the work of another.

Thank you again for your interest.

Sincerely,

Neil A. Armstrong
Professor of Aerospace Engineering

"DECLINE PARTICIPATION IN SUCH AFFAIRS"

January 12, 1972

Mr. Charles W. Reusing
President
Cincinnati Federal Savings and Loan Association
Cincinnati, Ohio

Dear Mr. Reusing:

Mr. Eyrich and Dr. Dennis have forwarded your request for my participation in the Cheviot-Westwood Kiwanis Club as Grand Marshall. I certainly appreciate your thoughtfulness in extending this kind invitation.

It has been my policy, however, to decline participation in such affairs. Such a position is consistent with my previous answers to similar requests for the Rose Bowl Parade, Sugar Bowl Parade, New York Thanksgiving Day Parade, etc.

Please accept my thanks for your interest and my very best wishes for the continued success of the "Harvest Rose Fair."

Sincerely,

Neil A. Armstrong
Professor of Aerospace Engineering

"I CAN SEE NO WAY OF ADDRESSING THIS SUBJECT"

January 24, 1972

Dr. Harold H. Wolf
Chairman, Colloquium Committee
The Ohio State University
Columbus, Ohio

Dear Dr. Wolf:

I have been delaying my response to your kind request because, frankly, I didn't know how to answer. I believe I could present a factual account of the use of drugs in space flight, but I am very reluctant to do so.

A presentation to "the entire university community" would seem particularly unwise. Even if the rationale were perfectly developed and stated, straight-forward as it is, I believe it would be misunderstood by some fraction of the audience.

Of more concern is the possible reporting of such a lecture. The general quality of news reporting today gives me no confidence that the subject matter would be interpreted accurately. A news account limited to the fact that I gave a lecture on the subject, with no reporting of the content whatever, would, in my opinion, do a disservice to those who are combatting the misuse of drugs.

In summary, Dr. Wolf, I can see now way of addressing this subject, even though I am perfectly comfortable with the history and decisions in that field over the past decade.

Thank you very much for the invitation, and best wishes for the success of this year's George Beecher Kauffman Lecture.

Sincerely,

Neil A. Armstrong
Professor of Aerospace Engineering

"THE CERTIFICATE IS HEREWITH RETURNED"

January 31, 1972

Mr. Henry N. Beard
Executive Editor
National Lampoon
Madison Avenue
New York, New York

Dear Mr. Beard:

Mr. Armstrong has asked me to reply to your January 19 letter notifying him that he has been awarded your Certificate of Achievement for "his part" in the production of the May 1971 cover of your magazine.

To accept the Certificate would imply that Mr. Armstrong knowingly contributed to the production of the May cover and/or endorses the content of the publication. Since neither is true, the Certificate is herewith returned.

Sincerely,

(Mrs.) Geneva Barnes
Secretary to Mr. Armstrong

The centerpiece of the May 1971 cover of the National Lampoon was the famous photo of Buzz Aldrin standing on the lunar surface, with Neil's reflection in his visor; of course, millions of people at the time, and still today, believe that the astronaut featured in the photo is Neil Armstrong. In the foreground of the cover design are characters from the Buck Rogers comic strip with the caption, "The Future: It Came from 1971." The cover also gave the title of one of the feature articles inside the issue: "Zero Gravity Sex Manual." The designer of the cover was Gray Morrow, famous for his National Lampoon cover designs.

"BRING THE CHILDREN TO SEE THE HORSE FARM"

February 27, 1972

Dear Mr. Armstrong—

John writes us from Chile that he has extended to you an invitation to be with us at Derby in Louisville. This is always the first Saturday in May—when Kentucky is at its prettiest.

We would be delighted to have you and Mrs. Armstrong as our guests at that time and at any time that you might like to bring the children to see the horse farm or take a ride on our "Belle of Louisville" on the Ohio.

I would have addressed this to Mrs. Armstrong if I had your home address.

With best wishes.

Cordially,

Dara G. Wood
(Mrs. Howard J.)
Louisville, Kentucky

Personal reply from Neil

March 28, 1972

Mrs. Howard J. Wood
Louisville, Kentucky 40207

Dear Mrs. Wood:

Thank you for your note and invitation to join you at the time of the Derby. I have standing invitations from several of my college friends who also live in the Louisville area. I only had one opportunity to see the race several years ago and as I understand it, the finish is still in contention.

My college friends have asked me not to come back until they have decided who owes whom how much on the basis of my last visit.

Please accept our very sincere thanks for your kind offer and our very best wishes for the future.

Sincerely,

Neil A. Armstrong
Professor of Aerospace Engineering

"LIKE TO HAVE AN APOLLO ASTRONAUT WITH US FOR THE INAUGURATION OF THIS GARAGE"

December 21, 1972

Dear Mr. Armstrong,

At the beginning of March next year we shall open a new garage in Lausanne (Switzerland) at the lake of Geneva.

As the name of our garage is Apollo, we would like to have an Apollo astronaut with us for the inauguration of this garage.

I have got your address from the Nasa. Do you think there is a possibility to have you as our guest? I would be very thankful to you for your coming and assure you that we would do our best to make your stay in Switzerland as pleasant as possible.

Hoping to hear from you soon we remain,

Your sincerely,

GARAGE APOLLO Henri Favre S A.
Hugo Decrauzat
Manager
Renens, Switzerland

Personal reply from Neil

January 3, 1973

Mr. Hugo Decrauzat
Manager
GARAGE APOLLO Henri Favre S.A.
Renens, Switzerland

Dear Mr. Decrauzat:

Thank you for your kind invitation to participate in the inauguration of your new garage. Regretfully, classes will be in session during that time period and I will, therefore, be unable to leave my university duties.

 Please accept my best wishes for the success of Garage Apollo.

Sincerely,

Neil A. Armstrong
Professor of Aerospace Engineering

"PLEASED TO MAKE AN EXCEPTION"

February 12, 1973

The Honorable J. W. Davis
Mayor, City of Huntsville
Huntsville, Alabama

Dear Mayor Davis:

I was most surprised to receive the certificate and key to the city. Normally, I do not accept honorary membership but am pleased to make an exception in the case of Huntsville.

 Your thoughtfulness is very much appreciated. Please convey my thanks to the appropriate officials of your city management.

Sincerely,

Neil A. Armstrong
Professor of Engineering

"MISS NUDE WORLD CONTEST"

May 29, 1973

Dear Neil:

Naked City (America's largest nudist resort) is a division of Air Check/
Videochex . . . and the site for the annual MISS NUDE WORLD
CONTEST (held this year on Saturday, June 30th), as well as the MISS
& MISTER NUDE AMERICA CONTESTS (to be held this year on
Saturday, August 4th)!

I wonder if you would honor us with your presence as a Judge for
either of these events? Other celebrity Judges in the past have included
Archie Campbell, Rocky Graziano, June Wilkinson, Johnnie Ray and
Robert Horton, as well as World famous authors, scientists and artists.
The Contest itself is held on Naked City's ultra-modern Sundial Stage,
and lasts approximately 90 minutes. Your round-trip expenses would
be fully paid, and either a Naked City Lincoln Limousine or helicopter
would transport you from Chicago's O'Hare or Midway airport directly
to Naked City.

I do hope you will accept my invitation, and wonder if you would
also be good enough to send me a personally autographed picture? I look
forward to hearing from you soon.

Most sincerely,

Dick Drost
President
Naked City, Inc.
Roselawn, Indiana

"MOON DAY"

July 20, 1973

Dear Mr. Armstrong,

In Feb. 1974 I will be doing a TV show about a project which I
have working for 5 years, a new National Holiday, "Moon Day,"
July 20, 1974.

If you support this project now, not 80 years from now, I have letter
from Senators and Congressmen who are for this idea.

Let's mark it now. I need your support today.

Best of luck in '74.

Richard Christmas
Lansing, Michigan

*An avid space enthusiast since boyhood, Richard Christmas, would write letters
to Congress for over forty years in an attempt to set aside a day dedicated to space
exploration. Although his letters did not result in a national "Space Day" as he
hoped, his letters led to ten Michigan cities, and sixteen cities in other states,
dedicating a day—and sometimes a week—to the study and celebration of space
exploration.*

"DOES NOT WISH TO GRANT THAT PERMISSION"

July 24, 1973

Miss Karen Astner
Permissions Editor
Hallmark Cards, Inc.
Kansas City, Missouri

Dear Miss Astner,

In reply to your letters of June 6 and July 17, requesting Professor Armstrong to sign applications for permission to reprint an excerpt from a "Life" magazine article, please be advised that Professor Armstrong does not wish to grant that permission.

Sincerely,

Ruta Bankovskis
Secretary to Professor Neil Armstrong

The article for which Hallmark Cards was asking reprint permission was "A Great Leap for Mankind," August 1, 1969, Vol. 67, pages 28–29. According to its Application for Permission to Reprint, Hallmark planned to publish the excerpt—315 words long—in a hardbound trade book, 64 pages long, entitled Great Moments of Our Time by Robert Aldace Wood. The excerpt would include Neil's famous first words on the Moon. For the world rights to this excerpt, Hallmark proposed paying Neil a fee of $21.

"I HOPE YOU CAN APPRECIATE MY POSITION ON THIS MATTER"

January 9, 1974

The Honorable C. L. Washburn
Assistant Secretary for Tourism
U.S. Department of Commerce
Washington, D.C.

Dear Mr. Secretary:

Thank you for inviting me to participate in your program to promote foreign tourism in the U.S.

This is certainly an important ingredient in the American economy, and I applaud your efforts toward its improvement. I ask, however, that I not be included in this particular article. Without going into individual reasons, I hope you can appreciate my position on this matter.

Best wishes for the success of your efforts in increasing the tourist business within the United States.

Sincerely,

Neil A. Armstrong
Professor of Aerospace Engineering

C. Langhorne Washburn (1918–2011) was a navy veteran of World War II, businessman (with Hiller Aircraft), and Republic Party leader who first became active in politics while working with several grassroots organizations toward the presidential election of Dwight D. Eisenhower in 1952. Although he kept his early career in the corporate world, he returned often to politics, among other things serving as director of finance for the Nelson Rockefeller for President Committee in 1964 and as director of finance for the Republican National Finance Committee from 1965 to 1969. Washburn then entered federal service, in 1970 becoming the assistant secretary of commerce for tourism for the Nixon Administration. Leaving government in 1977, he became vice president for Disney's Experimental Prototype Community of Tomorrow in Florida.

"I HAVE DECLINED PARTICIPATION IN 'FEATURE' AND 'FILLER' ARTICLES"

February 27, 1974

Mr. Joe Ziemba, Managing Editor
IPC Magazine
Scranton Publishing Company, Inc.
Chicago, Illinois

Dear Mr. Ziemba:

Thank you for your letter and invitation to be profiled in your spring or summer issue.

As a matter of policy, I have declined participation in "feature" and "filler" articles. I do occasionally author short pieces, and should I write such a piece in your field, I'll certainly keep your new magazine in mind for your possible interest.

Again, please accept my thanks for your kind invitation.

Sincerely,

Neil Armstrong
Professor of Aerospace Engineering

"WE AT *INVESTING* WOULD LIKE TO INTERVIEW YOU"

April 15, 1974

Dear Mr. Armstrong,

We at *Investing* would like to interview you on the subject of investing: how your ideas on how money should be managed.

The enclosed issue will give you a notion of the kind of interview we do: in January, it was with Art Buchwald; in February, Mario Andretti; in March, Art Linkletter. We've also interviewed Richard Tucker, Billie Jean King and Joe Namath.

Investing is distributed by some of the country's major brokerage houses to their larger clients. Our clients include Bache, Blyth Eastman Dillon, Hayden Stone, E. F. Hutton and Dean Witter. Our circulation is 250,000.

If you agree, we would come to Cincinnati at your convenience.

Sincerely,

Robert J. Cirino

Personal reply from Neil

April 17, 1974

Mr. Robert C. Cirino
Investing
Madison Avenue
New York, New York

Dear Mr. Cirino:

Thank you for your interest in including an interview article in a forth-coming issue of *Investing.*

As a government employee, and more recently as a university faculty member, I've had little contact with or participation in the investment community. My personal resources have always been so limited that I don't believe I could provide any interesting material for such an article.

Thanks, anyway, for your thoughtfulness in asking.

Sincerely,

Neil A. Armstrong
Professor of Aerospace Engineering

"KIND INVITATION TO ADDRESS THE CORPS OF CADETS"

April 24, 1974

Stephen Douglas Townes
Company D-4
US Corps of Cadets
West Point, New York

Dear Cadet Townes:

Your letter has been forwarded to me here at the University of Cincinnati after some delay.

I appreciate your kind invitation to address the Corps of Cadets, a privilege which I enjoyed three years ago.

Although I am a formal naval officer, and still can't take kindly your "Beat Navy" slogan, please don't conclude that is the reason for my declining your request. My duties here at the university and various government responsibilities limit my outside commitments severely making it impossible to participate in the number of events I might in other circumstances enjoy.

Thanks again for your thoughtfulness in asking.

Sincerely,

Neil A. Armstrong
Professor of Aerospace Engineering

"I HAVE NOT ACCEPTED INDIVIDUAL INTERVIEWS"

April 24, 1974

Mrs. Sallie G. Beck
Staff Coordinator
International Visitors Center, Inc.
Mercantile Library Building
Cincinnati, Ohio

Dear Mrs. Beck,

I have, for several years, maintained a policy of restricting my contact with the media to occasional press conferences. I have not accepted individual interviews, except when connected to significant news events.

I appreciate the difficulties that this presents to foreign correspondents, but I don't see how I can honor their requests while declining those of their American colleagues.

Sincerely,

Neil A. Armstrong
Professor of Aerospace Engineering

"WERE THE ASTRONAUTS HUMAN?"

August 26, 1974

Sir:

It was a pleasure talking with you briefly last week. As I mentioned at the time, I respect and understand your position on interviews. But you must understand that I'm a newspaperman, and all too often that means I have to badger, conjole, plead, pull my hair and stomp the floor to get a story. I'm paid to be a gadfly, which is why I'm buzzing around your head now.

I'm wondering if you would object to responding to a written question, replying either in writing or by telephone.

Actually, it's one major question, with many sides. It's one we touched on briefly over the telephone.

On one level, my question is simple: Were you frightened at the prospects of what might happen to you once you left earth? Did you prepare your wife for the eventuality that you might not return? Did you consider resigning? Did your experiences make you any more or less religious?

On a higher level, I'm really asking, "Were the astronauts human?"

It's not a ridiculous question, because they were not sold to us as human . . . and a selling job was essentially what it was. An extraterrestrial public relations build-up.

NASA had to sell the idea of spending billions of dollars on something of little tangible benefit. They sold it by making the astronauts, the training program, the entire goal of conquest of space, bigger than life.

It was all so damned noble, this beating those Russians and landing on the moon first. FIRST—that was the key word. To be second was to be beaten, to lose the race. FIRST was a national priority, and the astronauts were the personification of the dreams of a nation.

It must have been a crushing burden—but so much was spoken and written, and we hungry Americans ate it up so easily and quickly then begged for more, that it must have been so very easy to believe the press clippings. Did you begin to lose perspective? I would have.

Let's face it: you ARE different from me. You're not only driven by different things, but you've seen and experienced different things—things only a handful of humans have ever seen, experienced, lived.

But where does the difference stop? I'm quite sure that, given the chance, I would say 'No thanks' to a trip to the moon. Sure, I want to go—but I don't want to risk my life or to put in all the hard work that such a trip would require.

You stepped onto the moon the month after my wife and I were married—the week after we returned from our honeymoon, in fact. I cried that night. Honestly. And I think that one of the things that made it more appealing was you.

You were so . . . well, ordinary looking. That wasn't Randolph Scott or Cary Grant on that fuzzy screen. It was an everyday person. If he can walk on the moon, I thought, I can do whatever I want to do.

And I did. I became a writer. A moderately good one, too.

Yes, you were so HUMAN it was an inspiration—but on the other hand, you weren't. I'm still in awe of you, as I mentioned over the telephone. And I feel more than a little bit silly writing this letter to you.

But that's the effect the astronauts had on an entire generation. The Space Generation. I told my wife I had talked to you, and she wanted to know everything I said and everything you said and if you were pleasant and how your voice sounded and . . . well, you get the idea.

You must sometimes feel as though you're a ghost walking among the living: they want to touch you, but shy away from you. I envy you for a lot of reasons, but that isn't one of them.

I unwrapped a golf ball once. Cut open the cover and unwrapped all that rubbery stuff inside. I was both surprised and pleased to discover an ordinary black rubber ball at the center. It made me happy. I understood it; I knew how it worked. One of my friends told me there was a secret kind of water inside; another said there was a peanut; still another said there was a little bomb that would explode.

But today I know what's inside and that understanding gives me pleasure and a sense of accomplishment.

Aren't you tired of talking or thinking about the things I have brought up in the letter—surely it's something you've been over count-less times before.

But perhaps you would like the little rubber ball inside you to remain at your center forever. I can understand that.

I guess I'm asking you to unwrap yourself—if only partially. So I can understand. Also a lot of other people like me can understand. Then, maybe, so you can better understand.

It need not be lengthy or pretentious: a page, a paragraph, a three-minute phone call, a book, a word . . . whatever it takes.

With apologies for intruding upon you and with hopes of hearing from you soon, I remain,

Cordially,

Michael Willis
Lantana, Florida

Mr. Willis appended all the different phone numbers that Neil could call to reach him.

Personal reply from Neil

October 2, 1974

Dear Mr. Willis,

I have found your letter awaiting my return to the campus after being away for about a month.

I appreciate your candor in detailing your intentions and point of view. I will not change my position, however, . . . and when I do have something to say, I'll try to say it myself, rather than have someone else, with better articulation say it for me.

All the best,

Neil A. Armstrong
Professor of Aerospace Engineering

"CONTRIBUTED TO THE 'VALUE CONFUSION' OF CONTEMPORARY SOCIETY"

October 1, 1974

Mr. Paul B. Edwards
Director, Pacific Regional Office
United States Committee for United Nations Children's Fund
Pasadena, California

Dear Mr. Edwards:

I was very sorry to hear that you are going into the "celebrity auction" business. It seems to me to be one of those things that has contributed to the "value confusion" of contemporary society.

I have been an active UNICEF supporter. I do not, however, think that the end justifies the means.

I hope you will try to understand my position in this matter. I give my autograph freely to all who ask, but must return your cards unsigned.

Sincerely,

Neil A. Armstrong
Professor of Aerospace Engineering

Mr. Edwards was proposing to Neil that he sign a number of UNICEF greeting cards for the 1974 holiday season, which would be auctioned off to the highest bidder for the benefit of the hungry children of UNICEF.

"I CAN'T THINK OF ANY REASON I WOULD
WANT TO APPEAR ON A POSTER"

December 2, 1974

Vern Hammarlund, Inc.
Troy, Michigan

Dear Mr. Hammarlund:

You are certainly welcome to drop in anytime to discuss your proposed "Great Americans" poster project.

I don't want to imply encouragement. I can't think of any reason why I would want to appear on a poster.

Anyway, the door is open and phone line is clear (usually).

Sincerely,

Neil A. Armstrong
Professor of Aerospace Engineering

January 20, 1975

Dear Mr. Armstrong,

I decided to send some rough layouts to you rather than bring them to Cincinnati myself. I feel this will give you more time to consider my proposal, and the ideas indicated in the layouts.

Since you first raised the question of appearing on a poster, I have given a lot of thought to what your personal feelings might be, and would like to begin by first presenting my thoughts on this subject. As I said in my last letter, I honestly had not considered the fact that you might ask, "Why would I want to appear on a poster?" I'm sure you have other questions concerning the photograph, and I feel I can adequately answer them. However, when I've finished my proposal, and if you continue to wish to remain out of it, I will accept your decision and respect your privacy.

Recently, I attempted to gain a better insight into the post-lunar private lives of the astronauts. I visited the Kennedy Space Center in an attempt to better understand the entire space program. I also read Buzz Aldrin's book and large portions of another book sold at the space center. If anything, I've come out of it convinced that the astronauts are indeed, "Great Americans", and that Neil Armstrong is *right* for my poster.

It is very difficult to present my thoughts on this without sounding like I am trying to con you. My pitch is honest and straight forward, I hope you read it this way. I simply want to do another poster in my series of "Great" posters, with the idea being to tie-in to the 1976 bi-centennial. I want to do something on The Americana Theme. America badly needs a hero. If there is one underlying, universal feeling, that Americans have as a result of Watergate, it seems to be a feeling of distrust of public officials. The president has long been an American hero. Now we need a new hero. I've honestly got to say that I personally hold the astronauts in great esteem. I think I had the same lump in my throat and the same chills to see you step on the moon, as I would have at age 10. At age 43, I'd like to remind those who opposed the space program, that whether they liked it or not, it took one hell'ova guy to get into that capsule and let himself be blasted into space.

You might ask, "How could my poster possibly reach the millions of people necessary, to effect this problem one tiny bit?" It can't, but it's an important start. I'll print about 3,000 copies of the poster. I'll give them to art directors, writers, and creative directors, in key advertising agencies all over the country. This is a small segment of the population. It is often a jaded group of people, and outwardly at least, not taken to admitting to having a hero. I said "outwardly"—but these same people really are very human, typical, All American, and they help influence a lot of thinking in this country. This poster would come out at a time when all advertising people are deeply involved in projects for 1976.

I visualize a very human shot of you at dawn, standing at the end of main street U.S.A. A town like Milan, Ohio, where Thomas Edison was born—a town with a great square or main street would be nice. Possibly, Wapakoneta works for this. I've not been there yet, but I'll look it over. Possibly an early morning shot on the University of Cincinnati campus with you crossing the commons would be right. Possibly a shot in a classroom is better. The point is that these same advertising people recognize a "Great" job and that's what this is all about.

Possibly it is difficult for you to understand how I could want to put out so much effort for no monetary return. Are these posters not a "give away?" Yes they are. For me, the poster is therapeutic. It's a change from the day-by-day commercial world in which I work. This is my way of showing the advertising world that I'm still trying to be creative, that I want to compete, and that I like to conceive an idea and carry it out to completion.

There is another facet to this entire project I'll mention for whatever it is worth. I don't know if you still have commitments to Time-Life, Inc., or others, that would preclude your getting involved in advertisements or television commercials. I also do not know what your personal feelings are on the subject, but I did see your picture in an ad for General Time Corporation. If you are interested in commercials, there is always the chance that this poster could help in that direction by being on hand where those decisions are made.

To me, Mr. Armstrong, it all boils down to, "why not?" I simply want to take a picture of a "Great American." You would not be endorsing any product or any *person*. Your picture has been taken thousands of times. This time someone wants to print the picture and remind Americans that you did it first.

I would have my lawyer prepare, and give to you before we take any pictures, a letter guaranteeing that I would receive absolutely no income from the poster. I would also give you a substantial number of posters if you wanted them, for whatever use you wish. They will be designed so that my name can be easily cut off the bottom without effecting the picture of you. In my first letter, I stated that the project pays no money. I wanted to make it plain, that in no way could I compete with Time-Life, Inc. and other large industries. I did not want to embarrass you by offering you what we usually pay professional models for their services. Since my first letter, I have decided that I could, if you wish, make a donation of $100 to your favorite charity in your name. We often do this in our business to repay a person of substantial means, for his help in a particular project.

The layouts are rough but I believe you will get the general idea from them. I have also included a stat of a photograph that someone once did of Frank Lloyd Wright. It is a good indication of the type of strong, graphic form, that I would use in the poster of you. Layouts number 3

& 4 are indoor things, such as in your classroom and might have the feeling of the Wright photograph. I especially like the idea of a classroom since, because I feel a special closeness to this subject, having recently given a graduation address in Santa Barbara, California. I also will be lecturing at the Minneapolis College of Art and Design next summer and then again in California.

As a final note, Mr. Armstrong, I have included with this letter, a photograph of the John Kennedy memorial at Arlington. I shot it while on an assignment for United Airlines. I included it, because it shows another example of the type of rich color and light that I would use in the "—Great Americans" poster. I also sent it along, because I would like you to have it as a gesture of my appreciation for taking so much of your time up to this point. Mr. Armstrong, you still don't know me from Adam. I appreciate the time you have given to the consideration of my poster, and I'll accept your decision, whatever it may be. Many thanks.

Sincerely,

Vern Hammarlund
Troy, Michigan

Over the years Neil would receive all sorts of business deals and other schemes designed to take advantage of his fame and celebrity—he would say an unequivocal no to virtually every one of them. In this case, Vern Hammarlund of Troy, Michigan, wanted to publish a heroic poster of Armstrong as one of the great American heroes. It is actually quite surprising that Neil, in his first reply to Hammarlund, gave him his contact information, as Neil hardly ever did that. The likely reason was that Vern W. Hammarlund (1931–2015) was quite an accomplished photographer, known primarily for his advertising photography of the 1960s on behalf of the Detroit automobile industry.

Personal reply from Neil

February 5, 1975

Vern Hammarlund
Troy, Michigan

Dear Mr. Hammarlund:
You are a persuasive writer and I appreciate your candor. I must, however, decline your invitation. I simply prefer to not appear on a poster.

It was very thoughtful to send the Kennedy Memorial photograph. I can't keep it (because of its value) and I can't return it (crass), so I'll turn it over to our Art Department for others to study.

I send my best wishes for your continued success.

Sincerely,

Neil A. Armstrong
Professor of Aerospace Engineering

"LAUNCH THE BEST, BIGGEST AND MOST SUCCESSFUL OLDSMOBILE DEALERSHIP"

December 27, 1974

Dear Mr. Armstrong:

Just to plead my case, I have never hit a home run in the Big Leagues, never ran a touchdown against Green Bay, have yet to make a hole in one, was never offered a top spot in TV nor sang in Carnegie Hall, but what I am trying to do is a first in my profession.

To say that I am proud, is putting it mildly, but with your help and the help of many others, I hope to launch the Best, Biggest and Most Successful Oldsmobile Dealership in the Midwest. Please try to make my opening a success. Your presence would mean so much to me. This is the time Now! This is the Place! I would appreciate your help at this Hour. God bless you.

Sincerely,

GEORGE WHITE OLDSMOBILE, INC.
George W. White, Pres.
Cincinnati. Ohio

"MOST PLEASED TO AUTOGRAPH THE STAMPS FOR TENZING NORGAY"

January 9, 1975

Mr. Carlton S. Fernyak
President
The Mansfield Typewriter and Office Supply Company
Mansfield, Ohio

Dear Mr. Fernyak:

I would be most pleased to autograph the stamps for Tenzing Norgay. I had the pleasure of meeting him several years ago at an Explorers Club meeting in New York.

I appreciate your forwarding the picture and his autograph. Send the stamps at your convenience and I'll try to return them promptly.

We don't have the Bonanza [airplane] any more, but I do get a chance to fly enough to keep current.

Sincerely,

Neil A. Armstrong
Professor of Aerospace Engineering

Tenzing Norgay (1914–1986) was the Indian and Nepali Sherpa mountain-eer most famous for being one of the first two individuals known to reach the summit of Mount Everest, along with Edmund Hillary, on May 29, 1953. Neil Armstrong became a good friend of Hillary and a great admirer of Tenzing Norgay, thus agreeing to autograph stamps for him when he likely would not have done so for most other people. It is not known how Carlton S. Fernyak (1898–1987), a prominent Mansfield businessman, came to know either Norgay or Armstrong.

"FAMOUS PEOPLE'S EYE GLASSES MUSEUM"

January 11, 1975

Dear Mr. Armstrong,

We would like to add a pair of your eye glasses to the growing FAMOUS PEOPLE'S EYE GLASSES MUSEUM. It would be appreciated by us if you would send us a pair of your old glasses, regardless of their condition, along with a letter attesting to their authenticity.

Thank you,
Dr. M. J. Bagley
Director

Reply from Luanna J. Fisher, Neil's secretary

January 16, 1975

Dr. M. J. Bagley
Director
Famous People's Eye Glasses Museum
Henderson, Nevada

Dear Dr. Bagley,

Professor Armstrong has asked me to thank you for your letter of January 11, 1975.

Professor Armstrong does not wear glasses, and therefore, is unable to comply with your request.

Sincerely,

(Miss) Luanna J. Fisher
Secretary to Professor Neil A. Armstrong

"YOUR NAME WILL CARRY WEIGHT"

February 17, 1975

Dear Professor Armstrong,

I know you must be overwhelmed by requests of one kind or another, but I know how helpful you can be.

Enclosed is material about the black birds the Army is planning to destroy by chemicals, and alternatives to this plan which would be *better for man as well as the birds.*

I am hopeful you will feel it important to contact the Secretary of Defense, James Schlesinger, c/o of the Pentagon, Washington, D.C. 20301.

Your name will carry weight.

I am writing this as a representative of The League for Animal Welfare, a Cincinnati private humane society.

Sincerely,

Aimee B. Heilbronn (Mrs. Ralph)
Cincinnati, Ohio

P.S. The spraying of the blackbirds in Paducah, Kentucky, last night was an individual city project—lamentable, but something out of our hands. If we act promptly, we still hope to stop this at Fort Campbell, Kentucky, and Milan, Tennessee.

"RETURNING YOUR FLAG (UNDER SEPARATE COVER)"

February 27, 1975

Mr. Bob White
Baltimore, Maryland

Dear Mr. White:

Professor Armstrong has asked me to answer your letter. He is returning your flag (under separate cover) unsigned as it is his belief that marking on the U.S. flag constitutes defacement and is, therefore, illegal.

He would be pleased to sign any other non-philatelic items that you would provide in its place.

Sincerely,

(Miss) Luanna J. Fisher
Secretary to Professor Neil A. Armstrong

"MOTIVATE BY EXAMPLE RATHER THAN BY INDIVIDUAL DISCUSSION"

May 2, 1975

Mr. & Mrs. Dale V. Witt
Worthington, Ohio

Dear Mr. & Mrs. Witt:

Thank you for your letter on behalf of your son David.

The volume of mail received by Professor Armstrong precludes the possibility of returning individual motivational letters to young people. With the exception of his own students, practicality demands that he motivate by example rather than by individual discussion or correspondence. We hope you will try to understand our position in this matter.

If there is any way in which more information about the programs of this university would be of interest to your son, please let us know and we'll be very happy to forward the request to the proper office.

Sincerely,

(Miss) Luanna J. Fisher
Secretary to Professor Neil A. Armstrong

"WRITE A SMALL MESSAGE TO ELVIS PRESLEY"

June 19, 1975

Professor Armstrong!

I know you are very busy, but I sure would appreciate if you would take a moment of your time to write a small message to Elvis Presley?

After 21 years of Great Entertainment, he is entitle to a Salute! Ms. Joyce Gentry has selected me to write to Special People like yourself and ask them to write a message. As she is having a book published, which she will present to Elvis at his home in Tennessee this August. Several Governors, Mayors, Senators and Ball players have already sent messages and I sure would be very happy if the Greatest Hero in the World would Salute the Greatest Entertainer in the World!

Professor Armstrong, I am proud to have your picture in my Living room a Truly Great, American. Good Luck in Everything you do!

Thank you,

Bernadette Flynn
White Plains, New York

"APOLLO 11 COMMEMORATIVE BREAKFAST"

June 30, 1975

Dear Professor Armstrong:

The Sponsors and the Board of Advisors of the Apollo 11
Commemorative Breakfast are pleased to extend a special invitation to
you to attend and participate in this memorable event. We have always
felt that the launch of Apollo 11 should be memorialized as one of the
most significant events in the history of the Nation—if not the world.
There is need to preserve the legacy of our Space Program and to honor
the thousands of dedicated people who made it, and Apollo 11, the
greatest achievement of the century. The Charter Breakfast is dedicated
to this purpose. From its modest beginnings, we hope that it will grow
and flourish into an annual event of national significance.

The Charter Breakfast will be held at the Press Site on the John F.
Kennedy Space Center at 0830 hours on Wednesday, 16 July 1975.
Coffee and doughnuts will be served and short statements of commemoration will be made by key Apollo 11 participants like yourself. At T-3
minutes we will replay, by audio/visual means, the actual countdown
sequence of the launch—culminating with the lift-off of Apollo 11 at
0932 hours.

We would be honored if you would accept our invitation to attend
and speak at this event. As one of the Apollo 11 Astronauts, we are sure
that you have some deep feelings and fond memories that you would
like to share with the attendees. The actual program will last less than
one hour so we will not burden you with a long speech.

We sincerely hope that you will give our invitation your favorable
consideration and that you will be able to join us on this memorable occasion.

Sincerely,

R. P. Murkshe, Chairman, Canaveral Council of Technical Societies
George M. Megular, Vice-Chairman, Apollo-Soyuz Contractors

Personal reply from Neil

August 6, 1975

R. P. Murkshe, Chairman, Canaveral Council of Technical Societies,
Apollo 11 Commemoration Breakfast
Cape Canaveral, Florida

Dear Mr. Murkshe:

Your kind invitation to join you at the Apollo 11 Commemoration
breakfast was unfortunately mislaid and only recently came to my
attention.

Of course, I did not attend the launch and would not have been able
to participate in any case. I do appreciate the honor of the invitation and
am most pleased that Apollo 11 is remembered in this manner.

Sincerely,

Neil A. Armstrong
Professor of Aerospace Engineering

"NEVER DREAMED THAT SPACE FLIGHT WOULD BECOME A REALITY IN HIS LIFETIME"

July 1, 1975

Eric Felton
Edina, Minnesota

Dear Eric:

Professor Armstrong is away for the summer but I know the answer to your question.

When he was in school, he never dreamed that space flight would become a reality in his lifetime.

Probably you are unable to guess what might happen in yours.

Sincerely,

(Miss) Luanna J. Fisher
Secretary to Professor Neil A. Armstrong

"I REALIZE YOU COULD MAKE A SPEECH 7 DAYS A WEEK"

November 2, 1976

Dear Neil:

By now you have heard from Jim Greenwood regarding our fervent hope you can visit with and talk to the Ohio Pilots Association, a fine group of general aviation pilots right there in your own home state. I would not recommend you accept their invitation had I not been there at their annual meeting last year at the instigation of Mrs. Louise Timken, who sort of spear-heads the organization. Lou is one of the first ladies in aviation and is a board member with me on Wings of Hope, an aviation oriented charity. I'm sure you are acquainted with Lou.

The organization suggests a date next March, preferably March 11. They would jet down to Cincinnati, pick you up around 5 p.m., and have you back home from Canton that evening immediately following the dinner.

Neil, I realize you could make a speech 7 days a week, 365 days a year if you accepted all the invitations. Before I hit senior citizenship I was very much in demand on the burp circuit and speechmaking became anathema. But sometimes it was enjoyable and less of a duty than originally figured, and I do think the Canton appearance would be an enjoyable one for you as it was for me. One of the highlights of my visit was a reception at Lou's country estate with all the major industrialists of that area in attendance who have a deep interest in all things aeronautical.

Believe me, neither Jim Greenwood nor I would give you a bum steer. Rather, I suppose that nine times out of ten we'd recommend you decline burdensome invitations (altho they are all complimentary). With your CPT and general aviation background, with the occasion right there in your own back yard and with such great Americans as host, Jim and I know you'd find it a pleasant and rewarding evening, and your wife, too, if she could attend. And should you mention some of your early (CPT) days, I'd even fly up from Texas!

Please give it some thought.

Incidentally, after our dinner at Otto Pobanz's home last summer, I had dinner with Bill Long and gave him your regards. Not long afterwards we lost him. He died peacefully in his sleep. What a full life he had!

Sincerely,

George
GEORGE E. HADDAWAY
Dallas, Texas

George E. Haddaway (1909–1998) was one of Neil's long-time friends in the world of aviation. A prominent fixture in the north Texas aviation scene for decades as a pilot and aviation journalist, he became friends with many aviation legends around the world—not just Armstrong but also Jimmy Doolittle, Charles Lindbergh, and Wiley Post. Along the way he amassed a large collection of aviation memorabilia that would eventually be housed in the Doolittle Library at the University of Texas at Dallas. Haddaway also formed Wings of Hope, a charity that provided light aircraft as well as pilots and equipment for humanitarian relief to people in remote locations around the world. In 1975 Jim Greenwood (1920–2011) was senior vice president of Learjet. Previous to that, he had been the manager of press relations for Beech Aircraft and director of public affairs for the Federal Aviation Administration. He was recognized nationally as one of the premier authorities on the global general aviation industry. From 1972 to 1983 Otto Pobanz (1922–2015) was a prominent member of the executive board of the National Business Aviation Association. Pobanz himself was an aviation pioneer, beginning flight lessons at the age of sixteen in a Taylor Cub and amassing more than 1,000 hours by the time he was nineteen. Pobanz was one of only three commissioned officers under age twenty-one in the U.S. Navy during World War II. After the war he became active in the world of corporate aviation, becoming an executive pilot/captain for RCA in 1956 and chief pilot and flight operations manager for Federated Department Stores in 1957. Through the end of the century he was busy with numerous safety issues, serving on a number of aviation government/industry panels. William F. "Bill" Long (1894–1976) was another stalwart of the Texas aviation scene. A World War I aviator, Bill played a prominent role in the development of Dallas aviation as owner of the Dallas Aviation School and Air College and primary promoter of the development of Love Field.

Personal reply from Neil

November 12, 1976

Mr. George E. Haddaway
Dallas, Texas

Dear George:

How the Ohio Pilots Association coerced two non-Buckeyes to represent them is a mystery worthy of "Columbo" or "Ellery Queen."

You were most persuasive on their behalf, I must say. I wish I could be so eloquent in declining.

As Jimmie Doolittle and I are chairing the Lindbergh Memorial Fund Campaign, I would be obliged to spend my time to coerce them out of their billfolds (for a worthy cause). They would be sorry they asked me and I would be uncomfortable as well.

I do appreciate being asked by my fellow Ohio pilots and hope they'll understand.

I hope our paths cross again soon.

Sincerely,

Neil A. Armstrong
Professor of Aerospace Engineering & Applied Mechanics

"GO THROUGH QUITE A FEW ENSHRINEMENTS
BEFORE GETTING AROUND TO ME"

November 5, 1976

Mr. Kenneth W. Wisenbaugh, Staff Coordinator
Aviation Development Committee
Greater Cincinnati Chamber of Commerce
Cincinnati, Ohio

Dear Ken:

Many thanks for your kind recommendation to the Aviation
Hall of Fame.

   I personally hope, however, that they go through quite a few
enshrinements before getting around to me.

   All the best.

Sincerely,

Neil A. Armstrong
Professor of Aerospace Engineering & Applied Mechanics

"I HAVE NOT GENERALLY ALLOWED IT TO BE RAFFLED"

October 6, 1978

Dr. Neal P. Jeffries
Director
Center for Manufacturing Technology
Cincinnati, Ohio

Dear Neal:

I found your letter awaiting my return to the campus after the summer.

Although I have made my signature available for certain charities, I have not generally allowed it to be raffled at charitable auctions; however worthy.

I doubt that you would believe how many celebrity auctions there are every week!

I hope things are going well with you; hope our paths cross soon.

Sincerely,

Neil A. Armstrong
Professor of Aerospace Engineering & Applied Mechanics

"I ASK THAT YOU NOT WRITE THE PROFILE"

November 29, 1978

Mr. Donnie Wheeler
People Today Department
The Cincinnati Enquirer
Cincinnati, Ohio

Dear Mr. Wheeler:

Thank you for your thoughtful request for an interview. As you may know, I have, for some years, maintained a policy of restricting media contacts to those involving "hard" news. I believe this to be a reasonable position.

I acknowledge that you can, as you state, write feature articles without the permission of the subject. Public good and personal freedom are ever in opposition. I do not believe, however, that the Enquirer would choose to subjugate individual rights unless the public good were very well served as a result.

I ask that you not write the profile. By copy of this to Mr. Keating, I apprise your management of my wish. Should you feel you have a perspective that I've failed to consider, I'd be most pleased to know of it.

Sincerely,

Neil A. Armstrong
Professor of Aerospace Engineering & Applied Mechanics

A lawyer who had served as an assistant Ohio attorney general, William John Keating (b. 1927) became a Cincinnati Municipal Court judge and judge of the Ohio Court of Common Pleas in Hamilton County. From 1967 to 1970 he served on the Cincinnati City Council, before moving to the U.S. House of Representatives from 1971 to 1974, representing Ohio's 1st congressional district. Keating resigned from Congress to become president and CEO of the Cincinnati Enquirer. *Keating and Armstrong became good friends. Keating's brother, Charles (1923–2014), was a Cincinnati lawyer, real estate developer, banker, and financier, best known for his role in the savings and loan scandal of the late 1980s.*

Mr. Wheeler did not write his feature story on Armstrong. However, in its Sunday edition on July 18, 1999, long after Bill Keating had left the paper, the Enquirer *ran a long feature story entitled "Neil Armstrong, Reluctant Hero," by John Johnston, Saundra Amrhein, and Richelle Thompson.*

"MY OWN CONTRIBUTIONS WERE VERY MODEST"

August 14, 1979

John W. Vester, M.D.
Director
Clinical Support Laboratory
Good Samaritan Hospital
Cincinnati, Ohio

Dear Dr. Vester:

Thank you for conveying Mr. [Jaan] Kangilaski's interest in the parallel interests of Mr. [Charles] Lindbergh and myself.

Unfortunately, such a parallel is largely mythical. Although the work of an institution with which I was associated was excessively covered by the press, my own contributions were very modest and not worthy of special note. Candidly, any work I did was prompted by its engineering content and not by any medical connection.

I appreciate your personal attention to the matter and would be delighted if you could clarify it to Mr. Kangilaski.

Sincerely,

Neil A. Armstrong
Professor of Aerospace Engineering & Applied Mechanics

Jaan Kangilaski was editor of a news publication of the American College of Physicians known as The Forum. *In a phone conversation with Dr. Vester, he apparently declared that Neil's involvement in biomedical engineering theory and technology at the University of Cincinnati was strongly reminiscent of a similar involvement in medical matters by Charles Lindbergh. Editor Kangilaski asked Dr. Vester to contact Neil and persuade him to write an article drawing the historic parallels. Although he declined this invitation, Neil had gotten involved in a research institute affiliated with the University of Cincinnati that included George Rieveschl, a UC chemist famous for his invention of diphenhydramine (Benadryl), the first antihistamine, and Dr. Henry Heimlich, Cincinnati's famous inventor of the Heimlich maneuver, who practiced medicine at the local Jewish hospital. What brought the most publicity to this institute's work were reputed attempts to design a palm-sized artificial heart on the basis of the coolant pump that had been used in the Apollo space suit and to develop a portable artificial lung from a modified version of Apollo's PLSS (portable life support system). However, the project was actually only intended to investigate methods of reducing damage to blood cells while being pumped. The press extrapolated and created some confusion in this matter by exaggerating its ambitions and achievements.*

"I DISCOURAGE CELEBRITY AUCTIONEERS"

March 25, 1980

Mr. Guy C. Shafer
Group Vice President
Cole Industries, Inc.
Hartford, Connecticut

Dear Guy:

I was delighted to hear of your wife's interest in and contributions to the University of Hartford library.

While I am honored to be asked to provide one of my books, unfortunately I have never authored one. Both of my colleagues on Apollo 11, Mike Collins and Buzz Aldrin, did write books regarding their lives and experiences. Also, Crossfield, Bridgeman, Yeager, and Sevier are others who come to mind.

As a matter of fact, I provide autographs to all who ask without charge (philatelic items excepted). Consequently, I discourage celebrity auctioneers from implying that they are providing something that could not be obtained for nothing. I'm certain you can perceive the complexities that can ensue.

See you in Palm Springs.

Sincerely,

Neil A. Armstrong

Established in the 1960s, Cole Industries provided porosity-free aluminum for rotational molds. Neil apparently knew Guy C. Shafer.

"PERMISSION TO INCLUDE A RECORDED EXCERPT"

May 23, 1980

Dear Mr. Armstrong:

Attached is a script for a radio commercial we would like to produce for an educational seminar for EST.

We would like your permission to include a recorded excerpt of your speech at the time of your moon landing, as indicated in our script.

I would appreciate your advising me at your earliest possible convenience, if using this excerpt would be OK with you.

Sincerely,

Kathy Horan
Manager, Broadcast Business
Della Femina, Travisano & Partners Inc.
New York, New York

Attachments to this letter:

1. Vignette: (Kennedy making speech about the U.S. having a man on the moon within 10 years.)
2. Vignette: (SFX: of rocket noises with voice saying we have lift off . . .)
3. Vignette: (Voices from the moon's successful landing "one small step for a man a giant step for mankind".)
   V.O. [Voiceover] In a world where monumental acts are performed it is hard to comprehend that an individual can make a difference. That you and I matter at all.
   The possibility to create an atmosphere in which people realize that their lives really matter is undoubtedly the most profound opportunity available to anyone ever.
   If you knew that your life could matter, that you could make a difference, you would want that more than anything.

On July 18th and 19th at the Felt Forum in New York City there will be a filmed Seminar given by Werner Erhard that just might enhance the way you live your life. It is an opportunity to experience for yourself that you do indeed make a difference.

4. Vignette: (Personal experience of someone in which they did something special for a friend.)

V.O. "It's a you and me world." A Seminar given by EST, an Educational Corporation.

Personal reply from Neil

May 27, 1980

Ms. Kathy Horan
Manager, Broadcast Business
Della Femina, Travisano & Partners, Inc.
New York, New York

Dear Ms. Horan:

Thank you for your inquiry regarding the possible use of a recording of my voice for an EST radio commercial.

I have denied all such requests for such use by all organizations and companies as a standard policy. Due to the controversial nature of EST, I would violently object to the proposed usage.

Sincerely,

Neil A. Armstrong

"RATHER THAN THROUGH ANY THIRD PARTY"

December 19, 1980

Mr. Robert Daley
Robert Daley Productions, Inc.
The Burbank Studios
Burbank, California

Dear Mr. Daley:

I am delighted to know of your interest and success in the collection of historical documents and manuscripts.

I have delivered some of my papers to the Ohio Historical Society and have requests from a number of institutions, some with significant credibility. I will be considering their invitations in the future. I do expect to conduct such transfers directly to the chosen institution rather than through any third party.

Best wishes for your continued success.

Sincerely,

Neil A. Armstrong

Robert Daley was an American film producer and Clint Eastwood's producing partner from 1971 to 1980, working on such films as Play Misty for Me *(1971),* Dirty Harry *(1971),* Joe Kidd *(1972),* Hugh Plains Drifter *(1973),* Magnum Force *(1974),* Thunderbolt and Lightfoot *(1974),* The Eiger Sanction *(1975),* The Outlaw Josey Wales *(1976), and* Escape from Alcatraz *(1979). Prior to his work in film, he served as a producer for such television hits as* I Spy *(1965–1966),* The Invaders *(1967–1968), and* The Doris Day Show *(1968–1970).*

"I'VE CONCLUDED THAT THEY WILL BE AS MYSTIFIED AS I"

January 8, 1981

Mr. Al Reinert
FAM Productions
Houston, Texas

Dear Mr. Reinert:

I have taken excessively long in replying to your letter and returning the videotape, but I have considered the project extensively during that time.

I was favorably impressed by the footage selected and the quality of the preliminary editing. I was not, however, enthusiastic about the concept. I found the change of faces and voices confusing and the continuity difficult.

I recognized that I am at a disadvantage with my familiarity with the flights and crew members, but I tried diligently to view it from the perspective of the lay audience. I've concluded that they will be as mystified as I.

Some continuity could be achieved, perhaps, by using one single narrator on the voice over. I admit to some lack of confidence in the suggestion.

All in all, I do not believe that the addition of additional voice-over from an Armstrong interview will contribute to the effort and might just compound the confusion.

I do wish you well with the project.
All the best for 1981.

Sincerely,

Neil A. Armstrong

October 1, 1981

Mr. Al Reinert
FAM Productions
Houston, Texas

Dear Mr. Reinert:

As the tape player that is available to me has been out of service, it has taken some time to view the work-in-progress. I appreciate your patience in awaiting a reply.

When I watch the typical film in which I see a 747 mysteriously changed into a 707 in the next cut, I am offended. Similarly in this production when I am listening to the story of Apollo and am seeing Gemini footage, the film loses credibility.

The editing and voice over are intriguing and I approach accepting the changing crews and mis-matched voices. Still too much of a good thing becomes boring and the average viewer will find himself smothered in astronaut philosophy. The country music also seems overdone.

I have not changed my own position, but wish you well in the completion of the effort.

Sincerely,

Neil A. Armstrong

Mr. Reinert, indeed, did complete his effort and the product of his long efforts turned out much better than Neil anticipated. The production premiered in 1989 as For All Mankind, *a feature-length documentary that earned an Academy Award nomination. However, Neil stayed resolved and did not provide any voice-over for the film, unlike thirteen of the other Apollo astronauts (Lovell, Mattingly, Schweikart, Cernan, Collins, Conrad, Gordon, Bean, Swigert, Roosa, Irwin, Duke, and Schmitt). Neil did appear in the documentary as himself, of course, in footage of the Gemini VIII and Apollo 11 missions.*

Al Reinert went on to other film successes involving the space program. In 1996 he received an Oscar nomination for Best Writing, Screenplay Based on Material Previously Produced or Published for the highly popular Ron Howard/

Tom Hanks feature film Apollo 13. *For the miniseries* From the Earth to the Moon, *which aired on HBO originally between 1995 and 1998, Reinert wrote two episodes: Episode 4, "1968," and Episode 6, "Mare Tranquilitatis," the latter concerning Apollo 11. Reinert died in December 2018.*

"CONSENT TO THE USE OF YOUR NAME AND LIKENESS"

December 4, 1981

Dear Commander Armstrong:

Enclosed is a copy of a proposed advertisement which we are planning to run on behalf of our client, USAA [United Services Automobile Association].

We would appreciate very much if you would give us your consent to the use of your name and likeness in connection with our advertising. The advertising to be used by us will be the attached ad or very substantially similar. In no event will we make use of your name and likeness beyond five years from this date.

As consideration for your consent, we would like to make a contribution of $100 to your favorite charity.

If you will countersign a copy of this letter, it will constitute your agreement. Also, please designate your favorite charity.

Thank you.

Yours truly,

Stephen H. Plum
The Pitluk Group
San Antonio, Texas

The Pitluk Group—originally established in 1961 as the Pitluk Advertising Agency by Louis Harris Pitluk (1931–2016)—grew to become the largest advertising agency in San Antonio and one of the top five agencies in the state of Texas.

Personal reply from Neil

December 11, 1981

Mr. Stephen H. Plum
Senior Vice President
The Pitluk Group
San Antonio, Texas

Dear Mr. Plum:

I am in receipt of your request to consent to your use of my name and likeness in advertising copy for the USAA.

After reviewing the information provided, I have concluded that I will not grant consent. I have no wish to have my picture, name and quotes implying a non-existent association with an insurance company.

I do appreciate your bringing the opportunity to my attention.

Sincerely,

Neil A. Armstrong

"NEWSPAPER AD FOR EAGLE SNACKS"

November 15, 1983

Dear Mr. Armstrong:

Mr. John Uxa pointed out to us recently that you were the author of the phrase "The Eagle Has Landed" which we used in a newspaper ad for Eagle Snacks. He also suggested that we send you a sample of Eagle Snacks as a thank you. We agree with him. So under separate cover, we are sending you a sample of our famous Honey Roast Peanuts.

We hope you enjoy them.

Sincerely,

Allen Wm. Sherman
Vice President, Brand Management
Eagle Snacks, Inc.
St. Louis, Missouri

John "Jack" Uxa (1925–2017) was a lifelong resident of St. Louis. During World War II, he served in the Marine Corps Signal Company and was stationed on Saipan and Okinawa. After the war, he sold printing for many years before opening a successful dry cleaning store in suburban Warson Woods in 1969. The context for Mr. Uxa's giving William Sherman of Eagle Snacks the proper reference for "The Eagle Has Landed" is unknown.

Personal reply from Neil

December 9, 1983

Mr. Allen W. Sherman
Vice President, Brand Management
Eagle Snacks, Inc.
St. Louis, Missouri

Dear Mr. Sherman:

I recognize that such statements as you used in your advertising campaign are considered to be in the public domain. Their use in commercial activities is apparently legal. There are, however, many things that are legal but in poor taste.

Nevertheless, I appreciated your letter.

Sincerely,

Neil A. Armstrong

"AN ANECDOTE, RELATED BY JULIAN SCHEER"

January 3, 1985

Dear Mr. Armstrong:

We are considering publication of an anecdote, related by Julian Scheer, regarding a Walter Cronkite interview with you prior to the Apollo 11 launch.

Could you take a moment to review the anecdote for accuracy? A copy is attached.

Any assistance you can give me will be greatly appreciated. I can be reached toll-free at [number withheld] or collect [number withheld], between 8:30 and 4:00 Eastern Standard Time. Thank you for your time.

Cordially,

Ellen M. Bollert
Editorial Research
Reader's Digest
Pleasantville, New York

Attachment to this letter:
Former astronaut Neil Armstrong is the kind of man who speaks his mind. When Walter Cronkite interviewed the Apollo 11 crew just before the first moon landing, he asked what each man did to keep in shape. Mike Collins told Cronkite that he jogged about five miles a day. Buzz Aldrin said, "I not only run, but do an hour of isometric exercise."

"Well, Walter," Armstrong replied. "I believe that the good Lord gave us a finite number of heartbeats. I'm damned if I'm going to use mine up running up and down a street!"
—Julian Scheer, quoted by George Plimpton in *Esquire*

Esquire magazine published an article entitled "Neil Armstrong's First Words" in its December 1983 issue, pp. 113–18. Apparently Plimpton understood that most people believed that Armstrong had prepared his famous phrase about "one small step" epigram well in advance of the Apollo 11 launch. In this article,

Plimpton, apparently after an interview with Armstrong in 1983, "revealed" that Armstrong "had produced the lines on his own . . . and the words were composed not on the long trip up there, as had been supposed by most of his colleagues, nor beforehand but after the actual landing of Eagle on the moon's surface." Neil had explained to Plimpton that "I always knew there was a good chance of being able to return to Earth, but I thought the chances of a successful touchdown on the moon surface were about even money—fifty-fifty . . . Most people don't realize how difficult the mission was. So it didn't seem to me there was much point in thinking of something to say if we'd have to abort landing."

Personal reply from Neil

January 15, 1985

Ms. Ellen M. Bollert
Editorial Research
Reader's Digest
Pleasantville, New York

Dear Ms. Bollert:

I do indeed appreciate your contacting me regarding the quote attributed to me by Julian Scheer. Unfortunately, it is not factual.

The quote was first attributed to me in a LIFE magazine article after I, at a dinner party, had quoted someone else making the same remark. It has since been re-attributed to me frequently by other publications. It is usually portrayed as a "new" quote.

I have never made any such statement other than the one listed above. It certainly did not exist in the Cronkite interview (which I would expect to be in the CBS files).

I have been quoted so often that even some of my closest friends now believe it (and repeat it).

Again, I appreciate your confirming the authenticity.

Sincerely,

Neil A. Armstrong

"JOIN OUR 'HOMEGROWN HOOSIER' CAMPAIGN"

June 8, 1987

Dear Neil:

Perhaps you've heard or read about my recent disagreement with the folks at Merriam-Webster, Inc. Their recently published Webster's Third New International Dictionary defines "hoosier" as an "awkward, unhandy or unskilled person, especially an ignorant rustic."

I wrote them to complain. I told them "hoosier" should be defined as someone "quick, smart, resourceful, etc." They also said that if the word should come into common usage to mean "quick, smart, resourceful, etc." that they eventually would define it that way in their dictionary. That is our goal.

I've agreed to serve as honorary chairman of a campaign being launched by The Fort Wayne *News-Sentinel*, a Pulitzer Prize winning daily newspaper. We'd like you to join our "Homegrown Hoosier" campaign.

As a native-born Hoosier, we're asking you to help us change the meaning of the word. Sometime within the next few months we'd like you to use the word "hoosier" in some positive way: in an interview, a statement to the press, a song, article or story.

Once done, we'd like you to send published or taped proof of your public and positive use of "hoosier" to the *News-Sentinel*. The newspaper will publish them and see that the material is distributed as widely as possible. The *News-Sentinel* will also be contacting other daily newspapers in Indiana to join our campaign.

Send the material to Richard Battin, Assistant Managing Editor, The News-Sentinel, 600 W. Main St, Fort Wayne, IN 46802. For questions, call Richard at 219/461-8273.

Please join us in this noble cause. Wear the enclosed button as a symbol of your dedication.

And before you ask, the *News-Sentinel* paid for this mailing and the buttons and will pay any other expenses incurred during the campaign.

Sincerely,

Senator Dan Quayle

Reply from Vivian White, Neil's assistant

June 18, 1987

The Honorable Dan Quayle
c/o *The News-Sentinel*
P. O. Box 102
Fort Wayne, IN 46801

Dear Senator Quayle:

Mr. Armstrong has asked me to respond to your letter and to thank
you for offering him the opportunity to participate in the campaign to
change the definition of "hoosier." Since Mr. Armstrong is not a native-
born Hoosier, he believes that it would be inappropriate for him to
participate.

He sends best wishes for the success of the endeavor.

Sincerely,

Vivian White
Administrative Aide

"YOU TOOK RISKS AT MY EXPENSE"

November 9, 1987

Mr. Neil McAleer
New Cumberland, Pennsylvania

Dear Mr. McAleer,

Your article in *SPACE WORLD* regarding the 30th anniversary of
Sputnik illustrates the reason for my reluctance to collaborate in interviews that I cannot control.

Although my memory is suspect sometimes, I do not think I would have told you that I was flying the X-15 at the time of SPUTNIK as you reported. The X-15 rollout did not occur till December of 58, more than a year after SPUTNIK, and I did not fly it until 1960.

In your understandable desire to tighten a piece, you took risks at my expense. Others who know the facts will now suspect my memory is even worse than it is.

Sincerely,

Neil A. Armstrong

Neil McAleer came to write several books in the popular science and science fact categories, including The Cosmic Mind-Boggling Book, Body Almanac, The Mind-Boggling Universe, *and* Space Shuttle: The Renewed Promise. *His work* The Omni Space Almanac *won the 1988 Robert S. Ball Award for Outstanding Space Writing from the Aviation and Space Writers Association. He has also written a biography of Arthur C. Clarke (2012).*

I could not locate McAleer's response to Armstrong's letter, but Armstrong's subsequent response follows.

December 7, 1987

Mr. Neil McAleer
New Cumberland, Pennsylvania

Dear Mr. McAleer:

Thank you for your letter and explanation for the *SPACE WORLD* errors.

While I appreciate your offer to print a correction, I think little value would be gained by printing a correction. I have no desire to embarrass you or *SPACE WORLD*, and one more error in a vast sea will make little impact.

When a professional makes a slip of this size, it is easy to see why those with less rigor (including very successful authors) generate so much drivel.

Thanks for the offer.

Sincerely,

Neil Armstrong

"INCOMPATIBLE WITH MY OWN INTERESTS"

May 12, 1994

Ronni Heyman
Director of Public Relations
Mecanno Erector
New York, New York

Dear Ronni Heyman:

Thank you for your letter inviting me to attend the unveiling of an educational exhibition opening at the Hayden Planetarium on Tuesday, May 24, 1994.

I was impressed by the nature of this competition by students to be included in the exhibition in the display at the Planetarium. I suspect that it will be a most exciting project for all, whether their project is selected for inclusion or not.

After reviewing the information provided, however, I have concluded that I will decline. As a consumer products company, I know it would be important (and proper) for you to use this event for the promotion of your product line. The associated media activities that would be inevitable are incompatible with my own interests.

I appreciate your bringing the opportunity to my attention, and hope I find myself in New York during the exhibit time so I can enjoy the fun.

All good wishes.

Sincerely,

Neil A. Armstrong

Neil Armstrong was a big fan of Erector sets and played with one himself quite a bit as a boy. First created in England in 1898 and separately invented in the United States in 1913, Erector sets gave curious kids a chance to explore the principles of mechanical engineering by using metal nuts and bolts and miniature girders to build different sorts of structures, even dynamic ones.

"THROWING IN THE COMB AND SCISSORS TO BOOT"

November 23, 2005

Dear Neil—

I just saw this ad in a catalog of old political and historic memorabilia for sale. It looks like your enterprising barber and his client plan to make out on this—and they are throwing in the comb & scissors to boot!

Anyway, I'm sorry to see this indignity heaped upon you.

On a brighter side, I wish you and your lovely bride a blessed Thanksgiving and wonderful Christmas.

I hope all your 60 Minutes reviews have been positive; my wife has your new bio on my Christmas present list.

Best always,

Jim
James E. Rogan

In early 2005, the Lebanon, Ohio, barbershop that Armstrong had patronized for more than twenty years sold some of its famous client's locks for $3,000 to a Connecticut man who, according to Guinness World Records, had amassed the largest collection of hair from "historical celebrities." In a private conversation in the back of the shop, Neil asked his barber (Mark Sizemore) to either return the hair or donate the $3,000 to a charity of Armstrong's choosing. When neither result followed, Neil's attorney sent the barber a two-page letter, one *that referenced an Ohio law protecting the names of its celebrities. Instead of settling the matter quietly, the barber sent the letter to local media and refused to return the hair. According to the website for Paul Fraser Collectibles, not only Armstrong's hair but also the very scissors and comb that the barber allegedly used to cut it are for sale together for $38,611 (with free shipping). On the website, the company has also offered to sell individual pieces of Armstrong's hair for £399 (or $523) per each half-inch strand. Currently in 2019, one still sees strands of Neil's hair offered for sale on the internet, through secondary sales.*

Neil met Jim Rogan (b. 1957) when Rogan was a member of Congress, representing California's 27th District. A year before he wrote this short note to Armstrong, Rogan left a position in the Bush Administration (undersecretary of commerce for intellectual property and director of the U.S. Patent and Trademark Office) to join the law firm of Venable LLP, where he worked as a partner in its Southern California and Washington, D.C., offices. In July 2006, Rogan was appointed by California governor Arnold Schwarzenegger to serve as a judge on the Superior Court of California in Orange County, where he still serves.

7
THE PRINCIPLED CITIZEN

Compared to most other men, it seems correct to say, Armstrong lived by a handful of abiding principles—likely derived from his family upbringing, his dedication to the Boy Scott code and navy oath, but mostly, simply, from who Neil was as a very unique personality type. Neil had all the personal qualities that one would expect to find in a highly principled person: honesty, accountability, care and compassion, courage, fairness, gratitude, humility, loyalty, patience, and presence. But Neil's was a personality type whose conduct was very tightly joined to his sense of character. Even more than having the attitude that "it's not a question of what should I do but who should I be," for Neil it was matter of who he was—and who he wanted to be—a person very tightly in control at all times of what he should do. Being this sort of person, and believing this way, seems clearly at times to have constrained him from accepting different—potentially healthy—things in his life that his family, his friends, his fellow pilots and astronauts, his business associates, and his many thousands of regular fans and onlookers felt would in fact have been appropriate and generous acts on Neil's part, things he could have done that actually would have made his life easier.

No one in his life grew more frustrated with the steadfastness—even stubbornness—in keeping so tightly to his principles than his wife Janet. "He had so many requests for speeches and so many this and so many that—he didn't know where to start. He had to make decisions—and decision-making seemed to be especially difficult for him" the older he got.

"The man needed help," Janet felt. "I couldn't help him. He really didn't want me helping him. He didn't want to get angry with me, I suppose, or he didn't want me to get angry at him. That was probably smart on his part. Vivian White, his administrative assistant, used to get just beside herself. She just learned to go with the flow." Janet also tried to plan vacations for the two of them, but Neil couldn't commit—his schedule was always too busy. "I could not continue to live like that. He'd look at all sides of everything, and sometimes he'd discuss them—and I'd say, 'Just do it!' But he couldn't, or just didn't." It took a whole year even for Janet to get on his schedule to go away for a weekend, and Janet came to resent that greatly. "It really put the handwriting on the wall" for their ultimate separation in the early 1990s and divorce in 1994. "We had family. We had grandchildren. It was a long hard decision for me. It wasn't an easy thing to do—I cried for three years before I left." Janet had prolonged her decision because "the children were still there, the nest wasn't empty, there were still things going on. I always had hoped our life together would improve with time. I realized the personality. I just couldn't live with the personality anymore."[23]

In this chapter the reader will encounter many instances in which Armstrong's life of principled action—and even more so, nonaction—determined his participation in the civil life of his community, his state, his nation, and his world. If one could talk to many of the people who corresponded with him, as I have done over the past twenty years, one would hear a great deal of frustration and consternation with Neil for his apparent intransigence and stubbornness in his refusal to accept many seemingly innocent and positive invitations for his participation in different worthwhile causes and events. At the same time, readers will likely want to stand up and applaud Neil's firm and steady resolve to stay within his comfort zone and his sense of what is right and wrong, if just for him.

The chapter also reveals a great deal about Armstrong's political views. Some have said that Neil did not have a political bone in his body. "I don't think I would agree in the sense that I have beliefs, I participate in the process, and I vote my conscience. But what is true is that I am not in any way drawn to the political world."[24] He turned down opportunities to chair the Nixon reelection campaign in Ohio in 1972 and to run as a Republican against Democratic U.S. senator John Glenn in 1980. In terms of American political traditions, Armstrong always identified most strongly with the moderate roots of Jeffersonian Republicanism. "I tend to be more in favor of the states retaining their powers unless it's something only the federal government can do and it's in everyone's interest. I'm not persuaded that either of our current political parties is very right on the education issue. But it's not politic to express those views to anyone today. So I don't."[25]

"ACHIEVING A LASTING WORLD PEACE"

December 18, 1972

Dear Professor Armstrong:

I have undertaken a special study of viewpoints of leading world personalities in various fields of endeavors of achieving a lasting world peace.

The subject of a lasting peace is one of continuing concern for all of us. I am sure we all have thoughts on this subject and it seems to me an analysis of these opinions would serve a most meaningful purpose.

You could be most helpful to me in this project by giving me an expression of your views or formulas on the most effective means to achieve a peaceful world.

I am certain that you have some overall observations on the subject—such as your feeling on what we as a nation can do to contribute toward a peaceful world and what the community of nations might do towards this goal—and perhaps some deep-rooted thoughts or philosophical views which you wish to express.

I certainly would appreciate hearing from you on this most important of subjects, and you may be sure that I will greatly value your response. Also, as one who greatly respects your talents, I would be most appreciative if you could include an inscribed photograph of yourself in your reply.

With my highest regard and all my best wishes for the New Year, I am

Yours sincerely,

Seymour Halpern
Member of Congress

Personal reply from Neil

January 8, 1973

The Honorable Seymour Halpern
House of Representatives
Washington, D.C. 20515

Dear Mr. Halpern:

I am honored to be included among those from whom you seek viewpoints concerning the achievement of a lasting world peace.

Both my training and experience are in the world of technology, and it would be audacious of me to imply that my views in the fields of the humanities are as well reasoned as those of you who expend a great deal of thought on these matters.

At the risk of exposing the limitations of my understanding, I would offer the following observations: Man develops his thoughts and feelings into value judgments. The development of these value judgments yields ideals, the highest of man's attributes. This process is that which separates the man from the beast.

When these ideals are developed to the point of veneration, they also become worthy of defending (i.e., "Give me liberty or give me death").

The paradox seems to imply that man, in search for the improvement of his species, creates the conditions for war and his own self-destruction. Looking back through history, it is clear that the parade of armed conflicts over the centuries has had both terrifying short term effects and some beneficial long term results. It is not clear to me how these can be balanced.

If the human race is to continue one of four things seems to be required:

1. The right to defend one's values, with the consequence of armed conflict.
2. The establishment of a universal value system (It's not obvious how even to start toward this).
3. Giving up existing values, which would yield a regressive society.

4. Discovering an alternative method of defending values which does not require armed conflict.

The last alternative seems to hold the most promise, even if the technique is still beyond us. It seems the space program, in the past decade, provided a mechanism for competition without conflict between major powers, and it may have been effective. I would hope that there is something to be learned from that which could be valuable to your most important question.

Sincerely,

Neil A. Armstrong
Professor of Aerospace Engineering

"ARE YOU INVOLVED AND IF SO, HOW?"

March 15, 1974

Mr. Neil Armstrong
R.R. #2
Lebanon, Ohio 45036

Dear Neil:

Thank you for your very thorough and thoughtful letter in response to my request. You have "thrown the ball back into my court," and I shall respond. First of all, Neil, I was heartened to see that you are concerned and do have sensitive feelings about what is going on in our political and social worlds in which we live. Often, those who have something to say always find an excuse not to say it. So I suspect the question I have is, "Are you involved and if so, how?"

Certainly the two party system is important to the health of this nation. The party gives people of like persuasion a place to gather and a vehicle from which to launch their campaign; for, collectively they can be far more powerful than as only one lone voice. The party and its principles can mean different things to different people. I, for one, belong and work in the party because I need its collective voice in order to win my election, and I suspect that the party needs someone like me in order to voice its philosophy.

However, the party is not the savior, not the last word, and not the great hope. It is only the individuals in this country, when all is said and done that are important. Of course, overriding the importance of the individual is the health and welfare of the country; and, there is a direct correlation between the citizen and the state. If the individual is healthy, in other words his attitude, his thinking, and his approach to life are in order; for that citizen and citizens as a group will demand nothing less. Unfortunately, in the past century, we have seen a situation where the state has more influence over its citizens than the other way around. We have seen the citizen and his family raise the father or the breadwinner, who was once kind of his own domain, up to a level which we now call government. With this accession, the citizen and his family have passed on his authority and responsibilities to that government. Consequently, today we find the people looking toward government for guidance, leadership, and welfare. The citizen no longer is an authority, for he has subordinated this authority to government. That is the basis of the problem. There is a definite conflict between the individual's feeling of freedom, pride and responsibility, and that of the government's powerful hand, which is saying, "I know better. I will take care of you. Do it my way, but you must give up those personal qualities which you have traditionally reserved for yourself." That is what the fight is all about—an effort to restore those freedoms of pride, responsibility, authority and privacy in the individual by taking away the tremendous power of government. And here, our voices must be heard loud and strong. If it is not, I fear we will sink to the level described by George Orwell in 1984.

In order to be effective, each of us must operate in our own "modus operandi" in a way which is compatible with our own life style. I can certainly see your reluctance and the reasons why you have chosen not to speak out on behalf of one party, and I would be the last to pressure you into doing so. In your life's role, you have established yourself as a credible person, whom people respect. I feel that you would be doing a disservice to this country if you did not take advantage of that position, which you have worked and created yourself. You demand a forum, and you should certainly use it. You have a message, you have a concern, and you should express it; and, because of your credibility, others will listen. If you feel that the political world would undermine your position and this credibility, then certainly you should not follow that course. I do hope, however, Neil, that you will make your voice heard because as I said, if you and I and others do not, I am afraid that we will be in for some hard times.

Again, thank you for your kind response and for your thoughtful consideration. I hope that I have not bored you with this long letter, and I am hopeful that sometime when I am in Ohio, I may be able to stop by for some more in-depth discussion on this very serious subject. Give my best to Janet, and I hope that Susan and I can visit with you sometime in the future.

With best wishes,

Barry M. Goldwater, Jr.
Member of Congress
Washington, DC

Personal reply from Neil

March 27, 1974

The Honorable Barry M. Goldwater, Jr.
House of Representatives
Washington DC

Dear Barry:

I was most pleased to receive your thoughtful answer to my letter and particularly gratified for your understanding of my position.

Let me say that your letter shows extraordinary insights into the problems of the country, and I very much appreciate your sharing those views with me, particularly inasmuch as they coincide with my own.

I enjoyed your Goddard Luncheon address and felt you "dead eye aim" when reaching the remarks on scientific participation in the political process. I would agree that thoughtful, concerned citizens should make their voice heard. My own concern is the interpretation (or misinterpretation) of that voice by the press. I'm certain that you understand that concern.

You've honored me with your letters and request. I hope that I'll find a way to repay you. Please be assured that I'm sincere about that . . . and that you are Susan are always welcome here.

Sincerely,

Neil A. Armstrong
Professor of Aerospace Engineering

"ASKING YOU TO LEND YOUR NAME"

November 15, 1975

Dear Mr. Armstrong:

We are interested in projecting an entirely new dimension into the Democratic Presidential race—a new face from outside the field of politics—someone who is known by leaders all over the world, who has been nominated for the Nobel Peace Prize but who has never run for public office—someone who would promote a political dialogue on the concept of interdependence and would articulate the need for a dynamic new foreign policy based on the need for stronger global institutions.

We are writing to you with the hope that you would be willing to be one of 25 to 50 prominent Americans who would sign an ad in the New York Times requesting that Norman Cousins submit to a genuine draft and enter the Democratic primaries for the sole purpose of promoting such a dialogue. We are not asking for a contribution for that ad, although you may give one if you wish. Rather, we are asking you to lend your name as one of the prominent Americans who will be signing the ad. A preliminary draft of the copy of that ad is enclosed.

We recently commissioned a poll to determine the attitudes of opinion leaders on the question of a more effective world organization and the viability of a Cousins candidacy. The poll indicated that over 85% of opinion leaders in the United States favor the reform and strengthening of the United Nations and world institutions to solve our global problems, and that over 85% agreed it was important that Presidential candidates speak out on the need for a more effective world organization. The poll also indicated that Norman Cousins was the first choice of religious leaders and women leaders from among persons not in politics to enter the Presidential primaries to articulate these issues; the poll also showed he had broad support among editors, professionals and university presidents.

We have also sent out a small pilot mailing for charter membership contributions for the Draft Norman Cousins Committee. Contributions are beginning to pour in.

Norman Cousins has said that he will do anything for the cause of a just world order except make a damn fool of himself. We intend to demonstrate to him that there is sufficient support so that he would not make a fool of himself but rather would be commended by everyone for raising the level of the political dialogue and for forcing the other candidates to debate the fundamental issues of interdependence and the need for stronger global institutions.

We believe that in this year of post-Watergate, a draft of a person who has never before run for public office and who, if he consented to run, would do so only for the purpose of focusing national attention on the significant international issues, would have tremendous media impact. We believe it might well capture the imagination of the American people.

We recognize that getting Norman Cousins the Democratic nomination is a long shot—a very long shot; but we believe the effort will force a dialogue on the important issues and that alone will be worthwhile.

So we ask you: Would you please lend your name as one of the signers of the enclosed ad requesting that Norman Cousins submit to a draft and enter the Democratic Presidential Primaries?

Your early response by telephone and follow-up confirmation letter would be much appreciated.

Sincerely,

Walter Hoffmann

'76 DRAFT Norman Cousins Campaign
Wayne, New Jersey 07470

Personal reply from Neil

November 20, 1975

Walter Hoffmann
'76 Draft Norman Cousins Campaign
Wayne, New Jersey

Dear Mr. Hoffmann:

Thank you for your letter inviting me to participate in a campaign to initiate a drive to draft Mr. Norman Cousins as a presidential candidate. I have a great deal of respect and admiration for Mr. Cousins and can well understand the motivation of this initiative.

I, however, must decline. I have maintained a policy of political non-involvement and intend to continue in that manner for the immediate future. Any influence I might have is due primarily to the programs I was involved in, programs that were taxpayer supported under four administrations. I have concluded that it is most appropriate for me to refrain from taking public political positions.

I sincerely appreciate the honor of being asked to participate.

Sincerely,

Neil A. Armstrong
Professor of Aerospace Engineering

Norman Cousins (1915–1990) served for more than thirty years as the editor of the Saturday Review. *He authored a number of books on such topics as nuclear weapons controls, the United Nations, and world government. Under his direction the* Saturday Review *tackled such problems as environmental pollution, cigarette advertising, and violence in the performing arts. Walter F. Hoffman (1924–2014), a prominent civil rights attorney and advocate for world peace, was the national co-chairman of the Draft Norman Cousins for President*

Committee, a body of some fifty individuals who saw in Cousins the opportunity to promote world federalism. Cousins quickly chose not to pursue the candidacy.

"HOPE THAT YOU FEEL LIKE DOING THIS FOR ME"

July 14, 1976

Dear Neil:

It occurred to me this morning that maybe it was time for me to ask you to become more deeply involved in my political campaign.

As you are aware, the Senatorial election contest in New Mexico between Joe Montoya and myself has developed into a major test of the future political direction of this state and this Nation. The lines of choice could not be drawn more clearly as the enclosed news clipping shows. It is a choice between the old politics of personal financial gain versus a new politics of service through professional experience. Montoya is attempting to appeal to the instincts of short-term human desire while I am attempting to appeal to the people's needs and hopes for the future.

The results of the Senatorial Primary Elections in New Mexico and several public opinion surveys, including my own, show that the race is presently a dead heat. Support for my candidacy has increased from about 10% to about 40% in eight months of personal campaigning. This campaigning, the Watergate hearings, his IRS troubles, questions about his personal finances and his incumbency status, have decreased Montoya's support to about 40%.

As favorable as this looks for a challenger, four months prior to the election, there is no question that Montoya is still a very powerful opponent. He has political chits in from over forty continuous years in elective office. He has used a continuous flow of press releases on the disbursements of federal funds to create the illusion of great influence over his colleagues in the Senate. He has almost unlimited financial resources.

This summary is to prepare you for a request. It would be very useful to me if you would write personal letters to your professional colleagues in New Mexico asking that they support my candidacy with their vote and conversations. Membership lists in various professional societies are, of course, good general sources of names and addresses. I have enclosed some biographical and other materials which may be of use in composing your letter. If you prefer that I send you a suggested draft of a letter, please call or write and it will be done.

Well, I hope that you feel like doing this for me. If we do not begin to send men and women to the Congress who have the breadth of professional experience and the professional ability to get to the roots of our economic, research, development, health, defense and other major problems, then I see only increasing trouble in the rapidly changing third century of our national existence.

Sincerely,

Jack
Harrison "Jack" Schmitt
Candidate, United States Senator
Albuquerque, New Mexico

Personal reply from Neil

July 30, 1976

Mr. Harrison H. Schmitt
Albuquerque, New Mexico

Dear Jack:

I was happy to receive your letter and honored to be asked to help with your campaign. I certainly would be happy to see Senator Montoya ousted and would be overjoyed if you were the one to do it.

I do believe in everyone's right and obligation to participate in the political process and am delighted that you have chosen the senatorial candidate path. I know also that you have not seen me participate in any political activity. I have maintained a policy of noninvolvement in public politics and intend to continue that policy.

I am often approached by individuals from both parties, and have declined all including President Nixon and my friend Barry Goldwater, Jr. It is particularly difficult when the individual is someone whose friendship I value. I certainly hold you in the top of this category.

If there is any way I can be of assistance in a personal way, please feel free to call on me.

With every good wish and hope for your success, I am,

Sincerely,

Neil A. Armstrong
Professor of Aerospace Engineering

Harrrison "Jack" Schmitt, the lunar module pilot for Apollo 17, the last manned lunar landing (December 1972), defeated two-term Democratic incumbent Joseph Montoya in the November 1976 election to become the U.S. senator from the state of New Mexico; Schmitt's margin of victory was 57 percent to 42 percent. Though he served as the ranking Republican member of the Science, Technology, and Space Subcommittee, Schmitt's time in the U.S. Senate was short, one term; he was defeated in 1982 by New Mexico's attorney general, Jeff Bingaman, who attacked Schmitt for not paying enough attention to local matters. Bingaman's campaign slogan against the former astronaut (who resigned from NASA in 1975) asked, "What on Earth has he done for you lately?" Following his single term in the Senate, Schmitt became a consultant in business, geology, space, and public policy, and he went on to teach as an adjunct professor of engineering physics at the University of Wisconsin–Madison. A long-time advocate of returning to the Moon, in 1997 Schmitt cofounded Interlune-Intermars Initiative, Inc., to advance private sector acquisition and use of lunar resources.

"NO ONE COULD DELIVER AS EFFECTIVE A MESSAGE"

May 7, 1976

Mr. Neil A. Armstrong
Professor of Aerospace Engineering
798 Rhodes Building

University of Cincinnati
Cincinnati, Ohio 45221

Dear Neil;

I know Ken Wisenbaugh has been attempting to keep you informed of
the Chamber's Aviation Development activity, and we sincerely appreciate your interest.

A recent (and yet unannounced) development of great benefit
to Cincinnati is the agreement reached last week by the committee
with American Airlines which will restore early morning service to
Washington, D.C. The loss of early morning service several months
ago has been a real problem for many area business firms and especially
for the Environmental Protection Agency and National Institute of
Occupational Safety & Health personnel.

A difficulty the Chamber experiences is less than totally effective
communication with its member firms and the public about priorities
and programs. Too often people simply do not know of the day-in-
day-out effort of volunteers such as those who serve on the Aviation
Development Committee. We feel it is important that a better understanding be attempted of how things happen in a community like
Cincinnati and we've written a series of public service announcements
which WCPO is taping for us to use locally.

Neil, no one could deliver as effective a message regarding aviation
development in Cincinnati as you; so, I want to ask you first if you
would consider taping the following message for us:

"Hello, I'm Neil Armstrong. Greater Cincinnati Airport is among
the finest in the nation in the quality of air service it provides.
One reason is that the Greater Cincinnati Chamber of Commerce,
working directly with the Airport, the Civil Aeronautics Board and
airlines, has succeeded over the years in gaining major improvements in service—more non-stop flights and more conveniently
scheduled flights. Your Chamber of Commerce is working to keep
Greater Cincinnati a place where a good way of life keeps growing better."

Incidentally, so you know who is taping other messages, we have an OK from Peter Rose (sports), Oscar Robertson (minority business development), Eleftherios Karkadoulias (restored Tyler David Fountain, on downtown development), Betty Blake (riverfront development), Councilman Charles Taft (housing rehabilitation), Dr. Henry Heimlich (health-hospitals-medical/scientific research) and Police Chief Leistler (public safety).

Thanks for your consideration of this request. I hope sometime your schedule will permit you to visit the Chamber's offices; we do have one of the best organizations of its sort in the country and many of the things we do would be interesting and satisfying to you.

Sincerely,

Frank E. Smith, CCE
Executive Vice President
Greater Cincinnati Chamber of Commerce
120 West Fifth Street
Cincinnati, OH 45202

Personal reply from Neil

May 20, 1976

Mr. Frank E. Smith, CCE
Executive Vice President
Greater Cincinnati Chamber of Commerce
120 West Fifth Street
Cincinnati, Ohio 45202

Dear Mr. Smith:

Thank you for your letter explaining the public information program of the Chamber of Commerce. I can understand the desire to have a public awareness of the public service activities of the Chamber.

I know of the efforts of the Chamber, working hand in hand with the Board, to improve the quality of air service to our area. I do not feel, however, that it is appropriate for me to act as a spokesman for this work. It might be erroneously interpreted, by Kentuckians for example as a violation of the non-partisan character of a Board member's charge. I may be overly sensitive to this consideration, but I feel I must decline.

I appreciate the honor of being asked, and also appreciate the interest and dedication of the Chamber in our continued aviation progress.

Sincerely,

Neil A. Armstrong
Professor of Aerospace Engineering

From Neil to the Chamber chairmain

May 20, 1976

Mr. William Tillinghast
Chairman
Greater Cincinnati Chamber of Commerce
120 West Fifth Street
Cincinnati, OH 45202

Dear Bill:

I noticed that my name was included on the membership list of the Chamber Aviation Development Committee. In checking my files, I couldn't find any indication so I must have misled you or someone else at the chamber in personal conversation. If so, I apologize.

I am delighted to know of the committee's activities and progress, and hope I can continue to be aware of your work. I do believe, however, that I should restrict my active affiliation in this area to the Kenton County Airport Board, at least for the present.

I'll appreciate any help you can provide in correcting any misunderstandings that may remain.

Sincerely,

Neil A. Armstrong
Professor of Aerospace Engineering

From 1975 to 1979 Neil was an advisory member of the Kenton County Airport Board, after which he was appointed an honorary member for life. The Kenton County board owns and operates the Cincinnati/Northern Kentucky International Airport in Hebron, Kentucky, just across the Ohio River from Cincinnati.

"A POLICY OCCASIONALLY BROKEN . . . IS NO POLICY AT ALL"

October 26, 1976

The Honorable Bob Wilson
Member of Congress
Congress of the United States
House of Representatives
Rayburn Building
Washington, D.C.

Dear Mr. Wilson:

I found your collection of Apollo-related first day covers awaiting my return to the university after the summer holidays, accounting for the delay in this response.

I have, since leaving government service, maintained a policy of not signing first day covers (because of their commercial negotiability). I know, of course, that your collection does not fit in this category. A policy occasionally broken, however, is no policy at all. I know you recognize that it is much harder for me to write this letter than to sign the covers.

I have no other such prohibitions against autographing, and if you have any other items that would be appropriate, I'd be honored to oblige.

Best wishes for continued success.

Sincerely,

Neil A. Armstrong
Professor of Aerospace Engineering

*Robert Carlton "Bob" Wilson (1916—1999), a Republican, served as con-
gressman for California's 41st District from 1975 to 1981. The 41st District
represented Riverside County, located in the "Inland Empire" of Southern
California. I have not been able to locate the letter from Congressman Wilson.*

"REGARDING THE NIXON TAPES"

October 19, 1978

Mr. and Mrs. Neil Armstrong
798 Rhodes Hall
University of Cincinnati
Cincinnati, Ohio 45221

Dear Neil and Jan,

After a three year legal battle, the Supreme Court has ruled that
"The Nixon Tapes," despite having been played in open court in the
Watergate trial of Mitchell, Haldeman and Ehrlichman, et al., could not
be released to the public. This decision can be viewed as discriminatory
against the record and broadcast industries, since transcripts of the Tapes
are available in book form.

This Special Edition, limited to 200 copies, is an incomplete work.
Because of the Court's decision, it contains only the sections which were to
accompany the White House tapes. There is now one record where two had
been planned and silence instead of the voices of former government leaders.

I thought you might like a copy.

Very truly yours,

Michael Kapp
Producer-Director
Encino, California

Personal reply from Neil

October 31, 1978

Mr. Michael Kapp
Encino, CA 91316

Dear Mickey:

Many thanks for your note and the special edition of "The Nixon Tapes."

The public is mightily confused over the matter of confidentiality of records. In matters of criminal acts, one cannot very well require Nixon to turn his records over to the court and not require Farber of the N.Y. Times to do the same.

I admit some confusion in my own mind to the law and matters of privacy.

All the best.

Sincerely,

Neil A. Armstrong
Professor of Aerospace Engineering & Applied Mechanics

February 7, 1979

Dear Neil,

Kicking a dead horse is probably not the best way to spend one's time, however, your note of October 31 regarding the Nixon tapes has been plaguing me.

It's true that the public is confused over the matter of privacy, be it attorney-client, priest-parishioner, reporter-source, or public figure and press. The tapes we were seeking, however, were no longer Nixon's private property but the property of the court, thus we felt they would be treated in the same manner as transcripts.

I think it was Jefferson and those guys who wisely figured that a Supreme Court would be a good way to settle disputes over interpretations of the Constitution. I agree; I'm just miffed at that Court's unwillingness to render a clear-cut decision in our case by sending it back to the lower courts for three more years of litigation.

Best wishes.

Sincerely,

Michael Kapp
President
Warner Special Products
Burbank, California 91510

Michael Kapp was the president of Warner Special Products, Inc., a division of Warner Communications, Inc., and Warner Music Corp. Kapp came to know Armstrong through his production of To the Moon: The Dramatic Story of Man's Boldest Venture Told by the Voices of Those Who Achieved It, *a two-volume set of six vinyl records along with a hardcover book (in a pictorial cardboard slipcase), published in 1970. Michael was the son of David Kapp (1905–1976), a legendary record West Coast record producer who worked for Decca Records and RCA Victor before creating his own independent record label—Kapp Records—in 1954.*

As discussed in his letters to Armstrong, Michael Kapp in the mid-1970s endeavored to produce vinyl recordings for public sale of the "Nixon Tapes," the recordings of conversations between President Richard Nixon and members of his administration officials, White House staff, and Nixon family members made between 1971 and 1973 by a sound-activated taping system that had been installed in the Oval Office. The existence of these tapes came to light during the Watergate scandal of 1973 and 1974, with Nixon's refusal of a congressional subpoena to release the tapes, constituting one of the articles of impeachment against Nixon and Nixon's subsequent resignation on August 9, 1974. After Nixon's resignation, the federal government took control of all of his presidential records, including the tapes. For the next twenty years, until his death in 1994, Nixon was locked in legal battles over control of the tapes, with the former president arguing that the Presidential Recordings and Materials Preservation Act, passed following his resignation, was not only unconstitutional (in violating the principles of separation of powers and executive privilege) but also infringed his personal privacy rights and First Amendment right of association. (In his letter to Michael Kapp, Armstrong seems to be siding with Nixon's position.) In 1998, four years after Nixon's death, the courts ruled that some 820 hours of his tapes and forty-two million pages of documents were, in fact, his personal private property and had to be returned to his estate. In 2007, the Nixon Foundation gave control of the material—along with the other contents of what had been the privately operated Richard Nixon Library and Birthplace in Yorba Linda, California, to the National Aeronautics and Records Administration (NARA), with the facility renamed the Richard Nixon Presidential Library and Museum and becoming part of the federal presidential library system.

"SOME ACCOMPLISHMENTS ARE TRULY ENDURING"

June 7, 1979

Dr. Neil A. Armstrong
Department of Aerospace Engineering and Applied Mechanics
University of Cincinnati
Cincinnati, OH 45221

Dear Neil:

A short note of thanks. I appreciate your willingness to take time out from a busy schedule to participate in the Brown Commencement, also your graciousness in accommodating to the rushed demands of the ceremony.

You and I talked briefly about the changing attitudes of students over a short number of years. Especially for that reason I was thrilled—really thrilled—to see the seniors stand, to hear the sustained applause as President Swearer announced your name. It reminded me that some accomplishments are truly enduring, not only for their literal content but also for what they say about our corporate and individual spirit.

Madeline and I look forward to visiting with you in the near future. And some day—however distant—I hope we will be meeting to talk about next steps in the manned exploration of space.

Sincerely,

Tim
Thomas A. Mutch
Professor
Department of Geological Sciences
Brown University
Providence, Rhode Island 02912

Thomas A. "Tim" Mutch (1931–1980) was a distinguished American geologist and planetary scientist. In the 1970s he published two books—one about the geology of the Moon and the other the geology of Mars, the latter based largely on his role as the head of the surface photography team for NASA's Viking Project, which landed two unmanned spacecraft on the Red Planet: Viking 1 on July 20, 1976 (the U.S. bicentennial year and the seventh anniversary of the first Moon landing), and Viking 2 on September 3, 1976. A crater on Mars would be named in his honor, and the Viking 1 lander itself, after this death, would be renamed by NASA as the Thomas A. Mutch Memorial Station. Professor Mutch died in October 1980 after disappearing during his descent from atop the 23,409-foot Mount Nun in the Kashmir Himalayas.

"COLLABORATING WITH ME IN A SPECIFIC CONSULTING ASSIGNMENT"

June 14, 1979

Professor Neil A. Armstrong
Aerospace Engineering
University of Cincinnati
798 Rhodes Hall
Cincinnati, OH 45221

Dear Professor Armstrong,

We are fellow members of the National Academy of Engineering, so I am taking the liberty of writing to you to see if you might be interested in collaborating with me in a specific consulting assignment.

I have been asked by William G. Kuhns, Chairman, General Public Utilities Corporation, to form a small ad hoc committee of experts to make some recommendations to him concerning possible modifications to the man-machine interface problems in nuclear power plants.

This is, of course, in specific reference to the Three Mile Island incident and its rehabilitation and construction, but would presumably have wider applications in the industry also.

The effort contemplated is three or four top independent people from various fields working as a temporary committee. The total time involvement would probably approximate 15 to 20 days in the next six to nine months. I contemplate people from the disciplines of aerospace engineering, submarine operations, refinery and petrochemical engineering and possibly one other beside myself with a nuclear utility background.

General Public Utilities would pay your normal consulting fees under a letter of agreement if such arrangements were effectuated.

Would you please let me know either by phone or if more convenient by letter whether you are interested in this assignment and what your normal fee expectancy is.

I think this will prove to be an interesting assignment. It seems fairly clear that information presentation to the operator was at least one of the problems involved in the Three Mile Accident.

Faithfully yours,

Louis H. Roddis, Jr.
Consulting Engineer
Charleston, South Carolina 29401

Personal reply from Neil

June 21, 1979

Mr. Louis H. Roddis, Jr.
Consulting Engineer
Charleston, SC 29401

Dear Mr. Roddis:

Thank you for your letter inviting my participation in your study of man-machine problems in nuclear power plants.

I am a director of Cincinnati Gas and Electric Company. CG&E is in the final construction phase of the Zimmer Nuclear Power Plant of Moscow, Ohio.

Although I require an open mind to conduct the associated duties responsibly, I could be judged by others as failing the independence criteria your group would seem to require. I will, therefore, decline.

I send best wishes for the success of what could be a very important effort.

Sincerely,

Neil A. Armstrong
Professor of Aerospace Engineering & Applied Mechanics

"REAL TREAT TO THE PEOPLE OF WAPAKONETA"

August 21, 1979

Dear Mr. Armstrong,

I would like to take this opportunity to thank you for being here on July 21 and riding in the parade with your parents. It was a real treat to the people of Wapakoneta and the highlight of our three day celebration of the 10th anniversary of man's first step on the moon.

Our staff would be glad to arrange a visit to the museum for you and your family any evening you would like. We would be glad to have you go over the exhibits we have and would welcome your comments or suggestions.

Again, thank you for coming "home" and being a part of our anniversary celebration.

Sincerely,

John Zwez, Manager
Neil Armstrong Museum
The Ohio Historical Society, Inc.

Armstrong was not a big fan of the Wapakoneta museum in his name, as John Zwez, museum manager well knew. From the start (the museum held its grand opening on July 20, 1972, the third anniversary of the first Moon landing), Neil was not happy with how the entire museum project had come together, feeling that he should have been asked about it before plans to build it proceeded. "The policy I followed from the start," he stated in his interviews for First Man, *"had been that I neither encouraged nor prohibited the use of my name on public buildings, but I did not approve their use on any commercial or other nonpublic facility." If the organizing committee had asked him, "I'm sure I would have said okay, because it was in the town where my parents lived. Nevertheless, I would have been happier had they not used my name or, if they used my name, they would have used a different approach for the museum. I did try to support them in any way that I could by presenting them with such materials as I had available, either gifting or loaning items. From the outset I was uncomfortable because the museum was built as the 'Neil Armstrong Museum.' A number of people came to believe that it was my personal property and a business undertaking of mine. The Ohio Historical Society in Columbus was actually going to be overseeing the museum, and I told its director that I felt uncomfortable. I asked him as well as another member of the planning board if there was anything that could be done about the public image issue and to respond to me about what they thought. They said they would, but they did not." Armstrong's relationship with the museum and its leadership remained strained right up through Neil's death in 2012. In the mid-1990s, for example, came the issue of a picture postcard of Neil as an astronaut on sale in the museum gift shop. The image came from an official NASA photograph, taken when he was a federal government employee. For him it was a question of ownership. The rights to the picture belong to the people, the same visitors, Neil believed, who "think I own the place." The seal of the Ohio Historical Society was displayed inside the main door, but according to Neil, "it's so low profile that most people don't notice it." Eventually, Armstrong relented on the matter of the picture, granting museum manager John Zwez "my permission on a limited-time basis."*[26]

"YOUR DEDICATED LOYALTY CONTRIBUTED IMMEASURABLY"

October 17, 1979

Dear Neil:

Many thanks for the time and effort you so generously gave on behalf of the General Protestant Orphan Home as Hospitality Chairman. Your dedicated loyalty contributed immeasurably to the success of the 130th Annual Feast, and for this, we are indebted to you.

The proceeds from this year's Feast enable us to continue to serve the needs of children and their families in the area. Thank you for the part you played; it is appreciated.

Cordially yours,

Robert C. Schiff
Chairman
130th Annual Feast
Beech Acres—The General Protestant Orphan Home
6881 Beechmont Avenue
Cincinnati, Ohio 45230

"AS A RESPECTED AMERICAN"

June 4, 1983

Dear Mr. Armstrong,

On behalf of 24 distinguished Americans, I would like to ask your endorsement of an important new initiative, to resolve a concern of direct relevance to our daily lives. At stake is the purity of the water we drink and the air we breathe.

For, despite progress made since modern pollution laws were adopted in the 1970's, significant gaps in the statutory framework have allowed hazardous contaminants to escape control. For example:

—groundwater has been contaminated by synthetic organic chemicals, many of which are known or suspected to cause cancer, in 29 percent of 954 U.S. cities checked, yet there is no requirement for treatment or prevention of this risk to our drinking water supply;

—heavy metals are leaching into rural drinking water as one effect of acid rain—a growing problem, east and west, and not addressed under present law, which allowed sulfur oxide emissions (the primary cause) to increase by one million tons last year;

—federal standards limit the emission of only four hazardous air contaminants, out of more than 100 known or suspected to be harmful, including dioxin—despite measurements of likely chemical mutagens and carcinogens in urban air at 15–30 times median background levels.

A chance to close these statutory gaps—which exist regardless of who is EPA Administrator—and better protect health comes this year as Congress considers renewal of our air and water laws. The outcome of the debate is in doubt, however, due to a determined effort to weaken the law: to relax standards, delete funding, delay deadlines and otherwise impede our national clean-up goals.

To prevent this from happening, your help is needed. Specifically, would you be willing to join with other national leaders in business, labor and civic affairs, as an American for Clean Air and Water?

As a respected American, your endorsement of this bi-partisan effort will encourage our government to make a strong new commitment to contain toxic contaminants and ensure that our water is safe to drink, our air healthy to breathe.

As the same time, your expression of concern will help to resolve other important problems. In particular, gaining emission controls to reduce rainfall acidity—now averaging 40 times that of unpolluted rain—will serve also to protect thousands of lakes and streams endangered by its cumulative ill effects (plus save part of the $5 billion in annual damage estimated by the National Academy of Sciences and promote good relations with Canada).

Similarly, a greater resolve to keep clean the areas which are now clean can also halt the deterioration of visibility from our magnificent national parks—including the Grand Canyon, where views are obscured 100 days a year—and make our "purple mountains' majesty" easier to discern!

Our proposed approach is outlined in the attached position paper. It is consistent with goals of the national conservation coalitions active on these issues. Its emphasis on safe drinking water, however, and on providing an economic incentive as a supplement to needed regulation will allow us to approach Congress in our own right.

Your ideas and suggestions about how best to translate our concerns into results are most welcome. Please feel free to request background materials which might be useful. I would be pleased, of course, to discuss any aspect of this proposal.

We do hope that you will decide to join us as an American for Clean Air and Water.

Best Wishes,

Laurance Rockefeller

P.S.: We particularly hope that you can do this. It will make a very special difference.

Personal reply from Neil

July 5, 1983

Mr. Laurance Rockefeller
57 East 73rd Street
New York, New York 10021

Dear Mr. Rockefeller:

Your renewed invitation to join Americans for Clean Air and Water was as persuasive as it was unexpected. Nevertheless, as I told you in my earlier letter, I am not in a position to accept additional assignments at this time. The traumas associated with establishing a new business coupled with my existing civic and professional commitments preclude further affiliations.

I suspect that an American for Clean Air and Water would be a rare breed. A colleague (bio-chemist) from the university pointed out on a recent Earth Day symposium, however, that all substances are poison in sufficient quantities. Knowing that humans become ill from drinking excessive water or breathing excessive oxygen (however pure), I could understand his point.

It's not clear from the information provided how the organization intends to pursue its goals. I doubt that it believes any business can be effectively conducted with a 66 member committee! I do hope, however, that it is able to find an effective way to be helpful to the country.

With thanks and good wishes,

Neil A. Armstrong

Armstrong resigned from his faculty position at the University of Cincinnati in the autumn of 1979, to be effective the first day the first of the year 1980. For the past few years he had been growing more and more dissatisfied with university bureaucracy and its rules and regulations. He thought by moving into private enterprise he could take advantage of more of the opportunities that were coming his way in the corporate world, while doing better financially than he would ever be able to do at the university.

Neil's response to Laurance Rockefeller's invitation to become an "American for Clean Air and Water" says a great deal about his position on environmental issues. Clearly, it is not accurate to call him an environmental activist. Rather, his engineering background and conservative Republican values combined to make him quite cynical of many attitudes and political positions taken by the environmental movement in the United States and around the world.

Laurance S. Rockefeller (who died at age 94 in 2004) was the middle brother of the five prominent and philanthropic grandsons of John D. Rockefeller Jr (1874–1960). Unlike his older brother Nelson, who went into politics, Laurance concentrated on business and philanthropic work related to conservation, recreation, ecological concerns, and medical research.

"STANDING OVATION YOU RECEIVED"

March 7, 1985

Dear Prof. Armstrong:

The standing ovation you received when you walked into the Zoo Amphitheater conveys the feelings of the entire Cincinnati Zoo staff. We are happy that you and your wife could join us for one of our celebrity winter walks . . . and you know how much the crowd, even in the bitter cold, enjoyed your visit.

When the opportunity arises for you and your family to make another visit to the Zoo I will be glad to give you a personal behind-the-scenes tour.

Again, thanks.

Sincerely,

Edward J. Maruska
Director
Zoological Society of Cincinnati
3400 Vine Street
Cincinnati, Ohio 45220

"YOU MIGHT APPEAR MORE IN PUBLIC"

April 7, 1987

Mr. Neil A. Armstrong
777 Columbus Avenue/Suite 8
Lebanon, Ohio 45036

Dear Neil:

It has been a long time since we last spoke. However, our lunch in Paris left a lasting impression on me. You may recall that we discussed, if not the meaning of life, certainly a broad range of subjects related to engineering, aerospace and piloting. I was in Washington on business last week. During a meeting with the Air and Space Museum directors I again saw the X-15 and "Columbia." Naturally I thought of you.

It is my belief that you are one of history's most extraordinary people and certainly one of the extraordinary people I have met. As a result of having broken over seventy world ballooning records I have had the luck to meet enough exceptional people that I think I can have a valid opinion. Moreover to me you stand out as so much broader in outlook and more thoughtful than most astronauts and test pilots.

As such it seems a pity that you have chosen not to be a public figure in any way. In as far as I think I understand the reasons I naturally respect them. But with the passing of time your reasons may not be as valid as they were. This letter is simply to suggest that you might appear more in public, not to promote NASA, but simply because you are an inspiring person. Think of the young lives you might encourage and change for the better and the interest in math and science you could stimulate. It is surely clear that you are a far better role model than the athletes who seem to fill young minds.

I drafted this some time ago and hesitated to send it. I hope you will forgive me intruding now. But I am sure you receive many unsolicited suggestions and can easily deal with one more. My hope is that mine may just be the one to make the difference.

Respectfully yours:

Julian Nott
New York, NY 10021

Born in England in 1944, Julian Nott was a founder of the modern ballooning movement. As of 2017 he had broken seventy-nine world ballooning records and ninety-six British records, including exceeding 55,000 feet in a hot air balloon. Nott's records span many classes, including hot air, helium, super pressure, and combination balloons, and they encompass altitude, distance, and time aloft. Some of his world records have stood for more than thirty years. Nott and Armstrong met each other and became friends through membership in the American Institute of Aeronautics, Society of Experimental Test Pilots, and Explorers Club.

Personal reply from Neil

May 18, 1997

Mr. Julian Nott
New York, NY 10021

Dear Julian:

I was pleased to hear from you again after what seems like a long time. Please forgive the delay in this response, but I have been so busy giving public presentations that I am rather behind in my correspondence.

Since the time of your letter, I have given 3 presentations in Ireland (2 public), 1 to a regional Boy Scout meeting (public), 1 to a Community Club (semi-public), 1 to the Naval Museum Foundation [San Francisco] (public), 1 to the Naval Aviation Museum Symposium [Pensacola] (public), 1 to the Lindbergh Foundation [at the Air Force Academy] (public), and several to employees of businesses with which I am associated (private). Each of these were completely different from every other.

Over the last several years, I have given many public lectures in a number of countries, most recently (before Ireland) in Jordan, England, and India. In the enclosed recent clippings, the press reports, as always, my supposed reclusiveness. They learn this characterization from other newspaper articles. The primary reason for this view is that I have long maintained a policy of not granting individual interviews, and press conferences only when justified by newsworthy events.

I do not mind that my view of my public intercourse and that of the press do not coincide. It probably suits my purpose better than if we had the same opinion.

In any case, I am appreciative of your writing. If you have further thoughts, I would be delighted to know them.

Your reclusive friend,

Neil A. Armstrong

"LARGER KINDNESS THAN YOU REALIZE"

November 1, 1993

Neil Armstrong, Chairman
Computing Technology for Aviation
P.O. Box 436
Lebanon, Ohio 45036

Dear Mr. Armstrong:

You recently sent a signed photo to the Andrha Parish Neil Armstrong Team
here in India. I am enclosing some polaroids taken the day we presented the
picture. One photo shows the sons of the founder whose names indicate
how he views the accomplishments of you and your team aboard Apollo
Eleven. It is easy to find humor in the names but I am more impressed by
the fact that this small organization provides the only school available to
the children of a remote village, runs a home for indigent old people, and
is engaged in a moderate size Prawn farming project. The prawns provide
some money for the school but the farm is mostly a pioneering effort to
develop a new food source and a needed provider of employment.

Your gift of the photo was a larger kindness than you realize. Mr.
Nagendrarao had tears in his eyes when I presented it. Thank you for
helping to recognize love and charity at worl a long way from your home.

Very respectfully yours,

Boyd D. Sharp
Habitat for Humanity
c/o Maegabyte, 4/1 Arundelpet
Guntur 522-002, A. P. India

The address of the Society is:
R. Nagendrarao, President
The Andrha Pradesh Neil Armstrong Team
Siripudi (P.O.) Via Chandole
Guntur 522-311, A.P., India

"MY GRATITUDE TO YOU IS BOUNDLESS"

June 13, 1995

Dear Neil,

Thank you for writing me about the Sally Ride event.

I shall miss seeing you. But you have long since earned the right to do only those things you wish to do and in your own style. You have already given the public more of your life than most of us can even imagine. My gratitude to you is boundless.

Do keep in touch.

Affectionately,

Juanita Kreps
115 East Duke Building
Durham, NC 27708

Clara Juanita Morris Kreps (1921–2010) served from January 1977 to October 1979 as the Carter Administration's secretary of commerce. Like Armstrong, she was a person of "firsts": She was the first woman and first with a PhD in economics to hold that position. Before that, in 1972, she became the first director of the New York Stock Exchange. She knew Armstrong from their time together on the board at Chrysler, which for Neil went back to 1980. In 1987, Dr. Kreps became the first woman to win the Director of the Year award for the National Association of Corporate Directors. As with Neil, her efforts in the boardroom were seen as nothing less than the gold standard, believing, as he did, that it was the duty of a corporate board member to find the courage to raise even the most difficult issues for discussion, otherwise directors were falling down in their obligation to stockholders to make the most highly informed and intelligent decisions.

The Sally Ride event involved the presentation of the Von Braun Award to Dr. Sally K. Ride, the first American woman in space, by the Huntsville Chapter of the National Space Club.

The occasion of Dr. Krebs's letter seems to have been Neil's retirement from the Chrysler board, which came in 1995.

"NO ONE CONTRIBUTED MORE THAN YOU
DID TO THAT RESTORATION OF PRIDE"

June 20, 1995

Mr. Neil Armstrong
777 Columbus Avenue, Suite E
Lebanon, OH 45036

Dear Neil:

Thanks very much for taking time to send me a copy of your speech
to the 50th anniversary reunion of the Prisoners of War at Stalag Luft
3. I like the speech very much and I am sharing it with a couple of my
friends. One of them, a classmate from Bridgewater College days, just
joined me here as my Executive Assistant. I attended his separation
ceremony from the Marine Corps last week at the Norfolk Naval Base
and I was very touched to hear the information about his experiences
in Vietnam and other theaters. Another classmate I will send it to was a
fighter pilot in Vietnam and was badly injured by a missile attack on his
barracks. He has no movement in his body below the chest, but he has
learned how to drive a car, pilot an airplane, and function in an amazing
way. He is director of the Veterans Administration for Virginia.

It must have been a very moving experience to speak to these men. I
remember how emotional I was when I stood in the Arlington Cemetery
in Normandy a few years ago with two of my sons and saw the thousands of white crosses aligned so perfectly in the immaculate cemetery.
I was overcome with the realization that so many people had paid the
ultimate price for our freedom, that these young men, many of whom
were younger than my son standing beside me, had never made it home
and were buried three thousand miles from their country. It is important
that we remember and call attention to those sacrifices.

I appreciated your comments about the possibility that the space
program was a diversion from our Cold War focus. I had not thought of
it that way before but I think that may very well be true. In any event, it
helped generate excitement and pride in America during a period when
so much has led to cynicism, disappointment, and even loss of pride. No
one contributed more than you did to that restoration of pride.

Things are going well at the College. I am having a good time as a College President. I miss the practice of law but am still satisfied with what I am doing. I hope things are going well for you.

Sincerely,

Phillip C. Stone
President
Bridgewater College
Bridgewater, Virginia 22812

Stalag Luft III (officially Stammlager Luft III, or in English, "Main Camp, Air, III") was a World War II Luftwaffe-run prisoner of war (POW) camp, located in the German province of Silesia (now Żagań, Poland) some one hundred miles southeast of Berlin.

Phillip C. Stone served as president of Virginia's Bridgewater College from 1994 to 2010, after practicing law in Harrisburg, Pennsylvania, for two and a half decades. After five years of retirement when he worked with three of his children to start their own law firm, he came out of academic retirement in 2015 to take over as president of Sweet Briar College in Virginia, the historic women's college, until 2017. How Stone and Armstrong met is not known.

"LIMELIGHT SHOULD SPREAD A MUCH WIDER BEAM"

August 14, 2000

Dear Neil,

Your very kind letter of July 24 was forwarded to me in California. I am back here during the August recess sharing with my constituents the accomplishments of our Republican congressional majority. Also, as one who served on the House Judiciary Committee and was a House Manager in the impeachment trial of President Clinton, I am forced to spend an inordinate amount of time defending this action. My district is fairly heavily Democrat; in fact I represent most of the Hollywood studios. I was warned repeatedly that a vote for impeachment might well doom my future in the Congress. Those warnings may well prove correct in November. That is the difference between politics and the space program: politicians view "doom" as losing an election. I sometimes think we get a very twisted idea of what real risk means when we get stuck inside the Beltway.

You were very gracious to write. No thanks are necessary. The Apollo 11 congressional gold medal legislation I introduced is more than a labor of love: it is a long overdue recognition of the greatest accomplishment of the human spirit that occurred in my lifetime.

I hope you will permit me to speak candidly on the other issue you mentioned in your letter. I understand your sensitivity, and the sensitivity of your fellow crew members, to unnecessary publicity and (as you so eloquently stated) that the "limelight should spread a much wider beam." The dignity you have demonstrated these last few decades speaks volumes of your humility and recognition that this achievement was not a "three man show." I respect and honor those feelings. The Apollo 11 mission obviously was much more than the achievements of a ground crew at Mission Control. It was the culmination of every human dreamer that ever looked upward in thought and wonder.

Recognition of Apollo 11 in this legislation goes well beyond your tremendous personal accomplishment. It extends to each of those dreamers over the centuries who moved the ball toward the goal of manned space flight. It also reintroduces to a new generation of thinkers, dreamers, and (yes) political leaders the importance and potential of space exploration. If we lose sight of that goal, and allow our tremendous space accomplishments to be minimized, then future generations may look to the Mercury, Gemini, Apollo and succeeding missions as novel curiosities rather than epoch baby steps to expanding the limits of our earthly horizons. That would be the greatest of tragedies: all the sacrifice and labors of those who came before and after you would be in vain.

I serve with Members of Congress who are not old enough to remember the day the world held its breath and looked at the sky in a way never seen before or since. These Members who cast critical budget votes need to be taught, and their older colleagues reminded, of the importance of the space project and the vast promise it holds for the future.

None of the three Apollo 11 astronauts live in my district, so there are no votes for me to obtain by introducing this legislation. But there is a spirit that needs to be recaptured on Capitol Hill. And it is appropriate for the Congress, on behalf of the American people, to salute the crew who brought man's dream to life. In saluting the crew, Congress also salutes your predecessors, your contemporaries, and your heirs.

As you know, the bill has passed the House of Representatives. It is now being guided through the Senate by Senator Mike DeWine of Ohio. Once he obtains the sponsorship of 67 senators, it should move quickly. That probably sounds more complicated than it is: it involves simply having one's colleagues sign the petition of co-sponsorship. I obtained over 300 signatures in the House in just a few days, so when Congress resumes in early September Mike will do the same. I have talked to Mike and he has assured me he will walk the sponsor list around to his colleagues and obtain their signatures because he agrees with me that this is an important bill brought forward for an important reason.

Thank you again, Neil, for your thoughtful letter. As always, it is an honor to hear from you. I will continue to keep you posted on the progress of the bill. If you have any questions or additional comments, please do not hesitate to contact me.

With best personal regards from one honored to count you as a friend, I am

Sincerely,

James E. Rogan
Member of Congress
Washington, D.C. 20515

Armstrong, Collins, and Aldrin did, in fact, receive the Congressional Gold Medal but not until 2011, eleven years after Congressman Rogan first introduced the legislation to award the Apollo 11 astronauts the gold medal. On a Christmas card Rogan sent to Neil in 2000, he wrote: "I'm sorry my Apollo 11 Gold Medal bill never got out of the Senate. The Senator who promised to sponsor and move it never lifted a finger to help. By the time I found out, the time had run out. It was something I believed in very much. With or without a medal, you will always be my hero. Jim"

Personal reply from Neil

April 18, 2003

The Honorable James E. Rogan
Under Secretary and Director
U. S. Patents & Trademarks Office
Washington, D.C. 20231

Dear Jim:

Well, the National Inventors Hall of Fame is almost upon you and I remembered that I owed you a response to your last letter. Of course, you are free to read any part of the letter that you choose.

The Columbia disaster saddened everyone and reminded us that there is no progress without risk. Our charge is to maximize the former while minimizing the latter. So long as humans have an independent, creative and curious mind, we will continue to challenge the frontiers.

Thanks for your kind invitation. I will not forget. I send my very best wishes for good health and happiness.

Sincerely,

Neil A. Armstrong

April 30, 2003

Dear Neil—

Your very gracious letter of April 18 arrived today—just in time for me to read it at this Saturday's National Inventors Hall of Fame dinner. By the way, in this post-anthrax world of Washington, a two week delay in the mail is not bad. As of last week I was still getting Christmas cards!

You are so great to keep the dinner in mind & to send a few extra words for me to share with the group. The dinner will be taped & televised—we just never know when the next generation Von Braun may be listening in & maybe finding the great spark for life's ambition!

I know you prefer to travel quietly, and without fanfare. But if you are ever in DC and want to grab a few tacos at my favorite local Mexican restaurant, I'll do my best to insure our joint anonymity. At least people who recognize you greet you as a hero! If people remember me it is as a House of Representatives prosecutor in the Clinton Senate impeachment!

With my best personal regards to you & your lovely wife (whom I briefly met at Pete Conrad's funeral a few years ago), I remain

Sincerely,

Jim
James E. Rogan
Member of Congress (Ret.)
Arlington, VA 22202

Born in San Francisco in August 1957, James E. Rogan served as a member of the California state assembly before being elected as a Republican to Congress in 1997. As he mentioned in his letter to Armstrong, he was one of the lawyers appointed by the House of Representatives in 1998 to conduct the impeachment proceedings against President Bill Clinton. After losing his reelection bid in 2000, Rogan became the undersecretary of commerce and director of the United States Patent and Trademark Office for the Bush Administration (2001–2004). From 2006 to present, he has served as a judge for the superior court of California.

"WILL DO EVERYTHING TO ACCOMMODATE YOU"

July 9, 2009

E-mail to Neil Armstrong from Leslie Shumate, FOX News

Dear Mr. Armstrong,

FOX News would like to request a brief interview with you on either Saturday July 18th or Sunday July 19th (at possibly 4:40 pm Eastern). The segment, which would last between 4-5 minutes, would

be celebrating the anniversary of Apollo 11. We would love to have
you discuss your experience with Greg Jarrett and Julie Banderas on
America's News Headquarters. From what I understand, you are based in
Ohio, so we could arrange transportation to take you to a Columbus or
Cincinnati studio, where you could do a satellite interview. I know that
you are very busy and FOX News will do everything to accommodate
you and your schedule, should you be interested. Thank you.

*At the time she wrote Neil this email, Leslie Shumate had not yet graduated from
college and was a summer intern in the Fox News booking department, where
she booked guests for weekend programming, greeted guests as they arrived at the
New York studio, and escorted them throughout their taping. Her invitation to
Armstrong was professionally done, but, given the status of the man whom Fox
was after for an interview, one may be surprised that the task was given to a summer intern. It also seems clear from the email immediately below that Fox News
was likely unaware that three different people from Fox—Shumate, Clayton
Rawson, and Chris Wallace—were all individually and separately approaching
Armstrong about doing an interview on their network.*

Reply from Holly McVey, Neil's assistant

July 10, 2009

Email to Leslie Shumate, FOX News, from Holly McVey, Research
Assistant to Neil Armstrong

Thank you for your note. Mr. Armstrong has received invitations from
both Clayton Rawson and Chris Wallace at FOX and declined them
both. He normally restricts his interaction with the media to press
conferences and then only when he has some unique knowledge of
a newsworthy event. Thank you for bringing the opportunity to our
attention.

Holly McVey
Research Assistant to Mr. Armstrong

Personal reply from Neil to Clayton Rawson

July 10, 2009

Mr. Clayton Rawson
Director of Development
Fox News
New York, New York

Dear Mr. Rawson,

Thank you for your letter inviting me to participate in your forthcoming space commemoration program. I was delighted to learn that you remember warmly the Apollo programs that you produced.

I hope there is some sort of barrier between Fox and Fox News, but I cannot forget the airing of a dreadful special which, about a decade ago, charged Apollo as being a fraudulent program with our government and 400,000 Americans complicit in the crime. That was mildly abrasive, but when the program implied that we might have intentionally caused the Apollo 1 fire and the death of the crewmen to prevent them from confirming that the alleged fraud was true, I was mightily offended. One of those lost was my good friend and next door neighbor and it is impossible for me to forget.

Sincerely,

Neil Armstrong

On February 15, 2001, the Fox television network first aired its "documentary" Conspiracy Theory: Did We Land on the Moon? *The program did in fact claim that NASA faked the first landing in 1969 to win the Space Race against the Soviet Union. The principal credits for the program went to John Moffett (director, producer, writer) and Bruce M. Nash (executive producer, writer).*

"SPEAKING UP ON BEHALF OF AMERICA'S CONTINUED LEADERSHIP"

July 16, 2010

Mr. Armstrong,

I am a 5-time Shuttle flyer and now a NASA contractor. I manage Alliant Techsystem's (formerly Thiokol) site at the Marshall Space Flight Center, Huntsville, AL.

Thank you for your choice to go before Congress—not once but twice—to speak on behalf of a planned and reasoned approach to pursuing our nation's future in space exploration. Your testimony was balanced, thoughtful, honest, and effective.

Your testimony has reminded a generation of Americans that engineering success requires stark honesty and integrity. Politics and polarization of the different "camps" within the space community have made these rare attributes, and I thank you for bringing them into focus again.

All of us realize that you choose to support with care and thoughtfulness. That you chose this time and topic to make a rare public expression of your opinion only increased its impact.

Again, thank you for speaking up on behalf of America's continued leadership in the field of space exploration that you were so instrumental in establishing.

Sincerely,

Jim Halsell

A graduate of the U.S. Air Force Academy, James D. "Jim" Halsell Jr. (b. 1956) became an astronaut in July 1991. He logged over 1,250 hours in space as pilot on STS-65 (July 1994) and STS-74 (November 1995) and commander of STS-83 (April 1997), STS-94 (July 1997), and STS-101 (May 2000). From February to August 1998 he served as NASA's director of operations at the Yuri Gagarin Cosmonaut Training Center in Star City, Russia; upon his return to the United States, he served as manager, shuttle launch integration, followed by his role as Space Shuttle program manager for launch integration at the Kennedy Space Center (for the latter he was responsible for giving the "go for launch" on thirteen shuttle missions). After the Columbia accident on February 1, 2003, he led the NASA Return to Flight Planning Team. Halsell retired from NASA in November 2006.

"YOU WERE ON TARGET"

September 23, 2011

Dear Neil,

Thanks—so much—for the great visit. D.C. will never be the same. You were on target with your testimony & overly generous with your time. My friends who work for me—& got a visit & a picture—will never be the same.

Our office—our capital—our city—will never be the same. Thanks!!

Ralph M. Hall
Fourth District, Texas
Congress of the United States
House of Representatives
Washington, D. C. 20515

Ralph M. Hall served in the U.S. Congress from the Texas 4th District from 1981 to 2015. (Texas District 4 served an area of northeast Texas, along the Red River, northeast of the Dallas/Fort Worth metroplex.) Born in 1923, Hall was the oldest House member ever to cast a vote, at age 92. Hall was the chairman of the House Committee on Science, Space and Technology, before which Armstrong, joined by his fellow Apollo commanders Jim Lovell and Gene Cernan, testified in May 2010, speaking out against the Obama Administration's cancelation of NASA's Constellation program. Congressman Hall died on March 7, 2019.

"UNEXPECTED AND VERY MUCH APPRECIATED"

January 8, 2012

The Honorable Jean Schmidt
Member of Congress
Cincinnati, OH 45236

Dear Ms. Schmidt:

I arrived home after a skiing vacation in Colorado to find a large package awaiting. It came from you, of course, and was unexpected and very much appreciated.

The elegantly bound copy of your congratulatory statement in the House of Representatives, and the copy of the Congressional Record of November 16, 2011, on the occasion of the presentation of the Congressional Gold Medal to Senator Glenn and to the Apollo 11 crew was very kind and thoughtful of you.

The Congressional Gold Medal is the least frequently presented of the major honors of the United States government. And receiving the medal in the Rotunda of the Capitol Building from our Southwest Ohio colleague, Speaker John Boehner was an experience to be remembered for a lifetime.

I send my sincere thanks for your thoughtfulness and kindness along with my very best wishes for a happy, healthy, and productive 2012.

Yours very truly,

Neil A. Armstrong

On November 16, 2011, Neil received the Congressional Gold Medal, bestowed by the U.S. Congress to persons "who have performed an achievement that has an impact on American history and culture that is likely to be recognized as a major achievement in the recipient's field long after the achievement." At the same ceremony, held in the Capitol Rotunda, Mike Collins, Buzz Aldrin, and John Glenn also received the gold medals, the first recipient of which had been General George Washington in 1776.

Jean M. Schmidt served as Ohio's 2nd District representative to the U.S. Congress from 2005 to 2013. Ohio District 2 stretches along the Ohio River from the Hamilton Country suburbs of Cincinnati east to Scioto County and its country seat of Portsmouth.

APPENDIX

Secretaries, Assistants, and Administrative Aides for Neil Armstrong, 1969–2012

There was no way for Neil Armstrong to keep up with the heavy volume of mail he received over the years following Apollo 11 without considerable—weekly if not daily—assistance from secretaries or administrative aides. While he was still working for NASA, the agency did what it could to help Neil manage his mail by assigning a person or two from Public Affairs to open his mail, read each letter, and make a judgment as to whether it was a letter Neil would want, or need, to see. The assistant might also answer the letter, following guidance Neil had provided, always making sure the reply would satisfy him. Anything the assistant was not sure about, she most certainly ran before him, taking the side of caution, to make absolutely sure the mail was being handled exactly how Neil wanted. Naturally, as an assistant gained experience working on Neil's mail, she became more confident in her decisions and did not need to bother him as much in the triage of his correspondence.

Another thing the assistants did, particularly the assistants at NASA Public Affairs, was to keep stats on Neil's incoming and outgoing mail—that is, how many letters, cards, and so forth were received each month and how many replies the office sent on his behalf.

The following table provides the names of all of the women who helped Neil with his mail at NASA, at the University of Cincinnati, and through private arrangement with him once he went to work for himself after leaving the university in 1979. Of course, most people did not know when exactly Neil left NASA, arrived at Cincinnati, or left Cincinnati. Few knew

Years	Name	Organization
1969–1971	Shirley B. Weber	NASA, Office of Public Affairs
1971–1972	Geneva B. Barnes	NASA, Office of Public Affairs
1972–1973	Fern Lee Pickens	NASA, Office of Public Affairs
1971–1973	Ruta Bankovskis	University of Cincinnati
1974–1975	Luanna J. Fisher	University of Cincinnati
1976–1979	Elaine E. Moore	University of Cincinnati
1980–2003	Vivian White	Private (Lebanon office)
2003–2012	Holly McVey	Private

his working address, and even fewer knew his home address. So, letters kept arriving at NASA after he left, with the same being true after he left the university. For a short period after he left both institutions, his previous assistants continued to do what they could to help. After leaving the University of Cincinnati, Neil had to arrange for the administration of his mail and soon realized that handling his mail on his own was an impossible burden. In February of 1980 he rented a small office and post office box in Lebanon, Ohio, a little town north of Cincinnati, where the Armstrong family had moved to a farm after Neil resigned from NASA in 1971, and hired an administrative assistant—Vivian White—who managed his correspondence for more than twenty years. While he was at NASA, letters requiring translation were handled by government employees and contractors. At the University of Cincinnati, such letters were handled as best as could be from within the university community. The same was true when letters requiring translation arrived in the Lebanon post office for Vivian White to handle, as Vivian herself could not translate the letters.

While Neil was still with NASA and during his first couple of years at the University of Cincinnati, he also relied on these assistants to help him with his scheduling, bookings, and travel itineraries. Starting in 1974 he referred requests for nonacademic appearances—and there were a lot of them—to Mr. Thomas Stix, Stix and Gude, 30 Rockefeller Plaza, New York, New York 10020.

We know a great deal about how Vivian White handled her assignment with Armstrong, as she was interviewed at length for *First Man*; and from Vivian we also know a lot about Neil's policy for signing autographs.

Vivian White worked full-time for Neil for about ten years; after that she "cut back" to four-and-a-half days a week. According to Vivian, for the first twelve to fifteen years she worked for him, Armstrong would sign anything he was asked to sign, except a first-day cover. Then in about 1993 he discovered that his autographs were being sold over the Internet and that many of the signatures were forgeries. So he just quit signing. Still letters arrived in Lebanon saying, "I know Mr. Armstrong doesn't sign anymore, but would you ask him to make an exception for me?" After 1993, form letters under Vivian's signature went out in answer to 99 percent of the requests. In the few instances that Armstrong accepted the invitation, he composed and signed a personal letter. If he chose to answer someone's technical question, according to White, he would "write out his answer, I'd type it up and then put underneath it, 'Mr. Armstrong asked me to give you the following information,' and I signed it. We never answered personal questions. They were just too much an invasion of privacy." In Vivian's filing system, they would go into "File 11," the wastebasket.

It has been difficult to find biographical information about Armstrong's administrative aides, with the exception of Geneva B. Barnes and Vivian White. Thanks to a lengthy oral history interview conducted with Geneva Barnes on March 26, 1999, by historian Dr. Glen Swanson at NASA Headquarters as part of the NASA Johnson Space Center Oral History Project, we know a great deal about her. We also know quite a bit about Vivian White as she was interviewed at length as part of the research for *First Man*. Profiles of both Geneva and Vivian follow.

PROFILE OF GENEVA B. BARNES

Geneva B. Barnes—known to her friends as Gennie—was born on June 29, 1933, in Tahlequah, Oklahoma, so she was roughly three years younger than Neil.

During her senior year in high school in Tahlequah, Gennie was encouraged to enter into government service by her business administration teacher, who urged her to apply for a job with the Navy Department in Washington, D.C., knowing that a civilian recruiting officer would be visiting their school prior to Gennie's graduation, looking for talented stenographers. Along with her two best friends, she took the civil service test and received an appointment with the Office of Naval Material in Washington. It was the beginning of what turned out to be a forty-one-year career of federal service.

Gennie worked at the Office of Naval Material for four years, then accepted a position at the Pentagon in the Office of the Judge Advocate General (JAG). From there she moved to a position assisting the director of the Washington Regional Office of the Post Office Department (now the U.S. Postal Service). She was still working at the Post Office Department on February 20, 1962, the day Mercury astronaut John Glenn made America's first orbital flight in space. She left her office and stood in the rain watching Glenn in the company of Vice President Lyndon B. Johnson riding in a parade down Pennsylvania Avenue celebrating the historic mission. Afterward, she went back into her office, telephoned the NASA personnel office, and asked if it was hiring secretaries. Invited to apply, Gennie filled out an application with NASA the next morning and was, in her words, "on the payroll by 10 o'clock."

Her first job with NASA was as a secretary in the Office of Programs at NASA Headquarters. Then in 1963 she moved to the Office of Public Affairs, where she worked for the next eight years, handling many of the behind-the-scenes arrangements involved with NASA special events, including White House ceremonies and astronaut award ceremonies and appearances. During the Apollo missions, she assisted in protocol activities at Kennedy Space Center for four of the flights, including Apollo 11. One of the grandest and most personally rewarding events of her life was being part of the select group of support staff who accompanied the Apollo 11 astronauts and their wives on their Giant Step Presidential Goodwill Tour around the world from September 29 to November 5, 1969, visiting twenty-three countries in thirty-eight days. It was no vacation, however, as Gennie had worked hard with members of the State Department and President Nixon's White House staff setting up every detail of the project, including preparation of briefing materials and schedules.

As revealed in Armstrong's letters, Gennie served as Neil's public affairs assistant during his stint as deputy associate administrator for aeronautics in NASA's Office of Aeronautics and Space Technology. When Neil left NASA, she stayed on at OART until 1980, when she for a brief period worked as a management analyst for NASA's Office of Management Operations. But Gennie loved the astronauts and soon returned to a post as coordinator of astronaut appearances for the Office of Public Affairs. When she retired from NASA in 1994, she was handling the astronauts' appearances when they traveled internationally. After her retirement, Gennie spent many hours doing volunteer work at the White House in the e-mail section of the Presidential Mail Office.

In retirement Gennie lived in Capitol Heights, Maryland, just outside of the District of Columbia, due east of the nation's capitol.

PROFILE OF VIVIAN WHITE

Vivian White was born in 1921 to parents Archie and Lucille (Currey) Tartt in Kings Mills, Ohio, twenty-five miles northeast of Cincinnati. After graduating from Kings Mills High School in 1939, she began a secretarial position with Lou Romohr, a realtor in nearby Lebanon, Ohio, who later became the town's mayor. Vivian worked for Rohmer for twenty-eight years, during which time she earned her own real estate license, the first women in the history of Warren County, Ohio, to do so. In 1980, Lebanon's chamber of commerce honored her as Woman of the Year.

That same year Vivian went to work for Neil Armstrong, an assignment she handled with skill and dedication for the next twenty-three years, until 2003, from an office Neil rented in downtown Lebanon.

Vivian had two daughters, Lois and Margie, three grandchildren, four great-grandchildren, and two great-great-grandchildren, as well as several step-grandchildren and -great-grandchildren. She had one sibling, a brother named Archie Eugene Tartt. Besides working for Neil, Vivian enjoyed collecting antiques and playing bridge. She was active in the Town and Country Garden Club and Lebanon Council of Garden Club, organizations in which she held several leadership positions. She also was a long-time member of the Lebanon Business and Professional Women's Club.

Vivian White died on November 27, 2017, at the age of ninety-six.

NOTES

1. To this day it is a mystery as to whether Armstrong said "one small step for man" or, as he thought he said at the time, and certainly planned to say, "one small step for a man," which would have been grammatically correct. The world did not hear the "a," no question about that. In the rush of the moment, did Neil forget to say, or just not say, the "a"?

    In terms of memory, as Neil recollected, "I can't recapture it. For people who have listened to me for hours on the radio communication tapes, they know I left a lot of syllables out. It was not unusual for me to do that. I'm not particularly articulate. Perhaps it was a suppressed sound that didn't get picked up by the voice mike. As I have listened to it, it doesn't sound like there was time there for the word to be there. On the other hand, I think that reasonable people will realize that I didn't intentionally make an inane statement, and that certainly the 'a' was intended, because that's the only way the statement makes any sense. So I would hope that history would grant me leeway for dropping the syllable and understand that it was certainly intended, even if it wasn't said—although it actually might have been" (Armstrong to Hansen, September 19, 2003, p. 6 [see note 3, this chapter]).

    When asked how he prefers for historians to quote his statement, Neil told this author during interviews for the authorized biography, *First Man*, and answering only somewhat facetiously, "They can put it in parentheses" (Armstrong to Hansen, September 19, 2003, p. 8).

2. Quoted in Edwin E. Aldrin Jr. with Wayne Warga, *Return to Earth* (New York: Random House, 1973), p. 233.

3. Neil A. Armstrong to James R. Hansen, Cincinnati, Ohio, September 19, 2003, transcript p. 6. The cassette tape recordings of all of my interviews with Armstrong, as well as all of the verbatim transcripts of those interviews, are preserved in the Neil A. Armstrong papers collection in the Purdue University Archives and Special Collections, West Lafayette, Indiana. All interviews with Armstrong cited in these notes took place at Armstrong's home in suburban Cincinnati.

4. Willis Shapley to George C. Mueller, Office of Manned Space Flight, NASA Headquarters, April 1969, NASA Headquarters History Office, Washington, DC.

5. Armstrong to Hansen, September 19, 2003, pp. 8–9.

6. William H. Honan, "Le Mot Juste for the Moon," *Esquire*, July 1, 1969, accessed at www.classic.esquire.com.

7. For a fuller account of the Apollo 11 astronauts' reception aboard the USS *Hornet*, see James R. Hansen, *First Man: The Life of Neil A. Armstrong* (New York: Simon and Schuster, 2005 edition), pp. 555–57.

8. James A. Lovell Jr., 08:01:02:32 mission elapsed time, quoted in *First Man*, p. 562. There are two wonderful online resources by which one can follow the second-by-second course of all the Apollo missions. The first is the *Apollo Lunar Surface Journal*, a record of the lunar surface operations conducted by the six pairs of astronauts who landed on the Moon from July 1969 through December 1972. Edited by Eric M. Jones and Ken Glover, the *ALSJ* is a tremendous resource for anyone wanting to know what happened during the Apollo missions and why. The *ALSJ* includes a corrected transcript of all recorded conversations between the lunar surface crews and Mission Control in Houston. Interwoven into the journal are extensive commentaries not just by the highly informed editors but also by ten of the twelve astronauts who walked on the moon, including Neil Armstrong. The second such resource is the *Apollo Flight Journal*, which covers the eleven human flights of the Apollo program, including the five that did not land on the Moon, while at the same time relates those parts of the landing missions not taking place on the lunar surface. (One can find the words of Jim Lovell to the Apollo 11 crew just prior to Earth reentry in the *AFJ*.) Several editors have worked to produce the *AFJ*, notably Frank O'Brien, Tim Brandt, Lennie Waugh, Ken MacTaggart, Andrew Vignaux, Ian Roberts, Robin Wheeler, Rob McCray, Mike Jetzer, Alexandr Turhanov, and Ben Feist.

9. Armstrong to Hansen, September 19, 2003, transcript p. 35.

10. On Armstrong's relationship with the Boy Scouts, see *First Man*, pp. 31, 33, 37–39, 40, 43, 503, and 623–24.

11. Armstrong to Hansen, September 22, 2003, pp. 28–29.

12. Reverend Ralph Abernathy quoted in CBS Television Network, *10:56:20 PM, 7/20/69* (New York: Columbia Broadcasting System, 1970), pp. 15–16.

13. Soviet cosmonaut Gherman Titov quoted in Asif A. Siddiqi, *Challenge to Apollo: The Soviet Union and the Space Race, 1945–1974* (NASA SP-2000-4408, 2000), p. 667.

14. Soviet academician Leonid I. Sedov quoted in Siddiqi, *Challenge to Apollo*, p. 667.

15. Sergei Khrushchev quoted in Saswato R. Das, "The Moon Landing through Soviet Eyes: A Q&A with Sergei Khrushchev, Son of Former Premier Nikita Khrushchev," *Scientific American*, July 16, 2009, accessed at https://www.scientificamerican.com/article/apollo-moon-khrushchev/?redirect=1.

16. Ibid.

17. Armstrong to Hansen, Cincinnati, Ohio, September 22, 2003, p. 2.

18. Michael Collins, *Liftoff: The Story of America's Adventure in Space* (New York: Grove Press, 1988), p. 161.

19. Neil A. Armstrong, commencement speech, Wittenberg College, Ohio. Copy in the Neil A. Armstrong papers collection, Purdue University Archives and Special Collections, West Lafayette, Indiana. See also "Wittenberg Honors Armstrong," Akron (OH) *Beacon Journal*, November 24, 1969.

20. The *60 Minutes* episode on Neil Armstrong, "First Man," first aired on the CBS Television Network on November 6, 2005, the day before *First Man: The Life of Neil A. Armstrong* (New York: Simon and Schuster) hit the bookstores. This episode is available at https://www.cbsnews.com/news/neil-armstrongs-2005-interview-first-man/.

21. Robert Pearlman to James R. Hansen, Houston, Texas, January 20, 2019. Pearlman is founder and editor of collectSPACE, a website devoted to news and information concerning space exploration and space-related artifacts and memorabilia.

22. Janet Shearon Armstrong to James R. Hansen, Park City, Utah, September 11, 2004. These interviews and their transcriptions are not publicly available.

23. The relationship between Neil and Janet is covered quite fully in all three editions of *First Man: The Life of Neil A. Armstrong* (New York: Simon and Schuster). In the 2018 edition the final stage of their relationship, leading to their divorce in 1994, is discussed on pp. 364–66. These quotes are from

my private interviews with Janet Armstrong held at her home in Park City, Utah, on September 11, 2004.

24. Armstrong to Hansen, Cincinnati, Ohio, June 2, 2004, p. 4.

25. Ibid.

26. Armstrong to Hansen, Cincinnati, Ohio, September 22, 2003, p. 37. See also James R. Hansen, *First Man: The Life of Neil A. Armstrong* (New York: Simon and Schuster, 2005 edition), pp. 626–27.

ABOUT THE EDITOR

James R. Hansen is professor emeritus of history at Auburn University in Alabama. An expert in aerospace history and the history of science and technology, Hansen has published a dozen books and numerous articles covering a wide variety of topics, including the early days of aviation, the history of aerospace engineering, NASA, the Moon landings, the Space Shuttle program, and China's role in space. In 1995 NASA nominated his book *Spaceflight Revolution* for a Pulitzer Prize. His book *First Man*, which is the only authorized biography of Neil Armstrong, spent three weeks as a *New York Times* Best Seller in 2005 and 2018 and garnered a number of major book awards. Translations of *First Man* have been published in more than twenty languages. A Universal Studios film adaptation of the book hit the silver screen in October 2018, with Academy Award winner (*La La Land*) Damien Chazelle directing the film and actor Ryan Gosling starring as Armstrong. Hansen served as coproducer for the film.

Hansen began his career in aerospace history while serving in the early 1980s as historian-in-residence at NASA Langley Research Center in Hampton, Virginia. Over the years he has served on a number of important advisory boards and panels, including the Research Advisory Board for the National Air and Space Museum, Editorial Advisory Board for the Smithsonian Institution Press, and Advisory Board for the Archives of Aerospace Exploration at Virginia Polytechnic Institute and State University. He also is a past vice president of the Virginia Air and Space

Museum in Hampton, Virginia. For the past ten years he has served on the National Air and Space Museum Trophy Selection Board. His experience as an academic and public speaker has been wide-ranging both topically and geographically; he frequently serves as keynote speaker, panelist, and lecturer on a wide variety of topics in the history of science and technology.

A native of Fort Wayne, Indiana, Hansen has been married to Margaret Miller Hansen, also a Fort Wayne native, since 1976. They reside in Auburn, Alabama, and have two children and four grandchildren.